Peter D Riley

Cambridge

che **t**

checkp●int
Science

3

HODDER
EDUCATION
AN HACHETTE UK COMPANY

Titles in this series

Cambridge Checkpoint Science Student's Book 1	978 1444 12603 7
Cambridge Checkpoint Science Teacher's Resource Book 1	978 1444 14380 5
Cambridge Checkpoint Science Workbook 1	978 1444 18346 7
Cambridge Checkpoint Science Student's Book 2	978 1444 14375 1
Cambridge Checkpoint Science Teacher's Resource Book 2	978 1444 14381 2
Cambridge Checkpoint Science Workbook 2	978 1444 18348 1
Cambridge Checkpoint Science Student's Book 3	978 1444 14378 2
Cambridge Checkpoint Science Teacher's Resource Book 3	978 1444 14382 9
Cambridge Checkpoint Science Workbook 3	978 1444 18350 4

To Anita

Hachette UK's policy is to use papers that are natural, renewable and recyclable products and made from wood grown in well-managed forests and other controlled sources. The logging and manufacturing processes are expected to conform to the environmental regulations of the country of origin.

Orders: please contact Hachette UK Distribution, Hely Hutchinson Centre, Milton Road, Didcot, Oxfordshire, OX11 7HH. Telephone: +44(0)1235 827827. Email education@hachette.co.uk. Lines are open from 9.a.m. to 5.p.m., Monday to Friday. You can also order through our website www.hoddereducation.com

Cover photo © Gusto images/Science Photo Library
Illustrations by Barking Dog and Pantek Media.
Typeset in 12/14pt ITC Garamond Light by Pantek Media, Maidstone, Kent
Printed and bound by CPI Group (UK) Ltd, Croydon, CR0 4YY

A catalogue record for this title is available from the British Library

ISBN 978 1444 14378 2

Cambridge checkpoint

NEW EDITION

checkpoint Science 3

Contents

Preface

To the student

Can you remember when you started this science course two years ago? We began by looking at the activities of different scientists – marine zoologists, chemical engineers and astronomers, and then went on to learn about the Bunsen burner, laboratory equipment, the microscope and how to use them safely. You started developing scientific enquiry skills and building up your scientific vocabulary by using words such as *variable* and *data* as you made your investigations and began an exploration of the three main sciences – biology, chemistry and physics.

Last year you took your scientific enquiry skills further by considering how to think creatively on your own, or with others in brainstorming sessions, and to make simple calculations on the data you collected. At the same time you made closer examinations of the body to find out how organs, such as those in the digestive and circulatory systems, worked, you learnt about the discovery of elements and how their atoms combine in chemical reactions to form compounds and discovered that the particle theory could be used to explain how sound travels but that light is a completely different phenomenon which can be considered either as an electromagnetic wave or a particle of energy called a photon.

This year, as you take your scientific enquiry skills still further, we begin by considering 'What makes a scientist?' We will look at the stages we all go through as we grow up for signs of developing skills which are useful in the study of science. This is further explored throughout the book by looking at scientists from the past and present to see how their lives in science developed. Among the diversity of topics we study this year are ecosystems and how living things adapt to their environment, the structure of atoms and the rates of reactivity of chemical substances and we compare electrostatics with the flow of electricity in a circuit.

No study of science today would be complete without considering the environment and our survival in it. Last year we looked at how medical science is used to fight disease. This year we look at how human activity has adversely affected the environment and what we must do to correct it and we look at our energy needs and what we must do to live within the limits of our planet's resources. All these areas of study will need scientists in the future; perhaps you might be one of them.

Cambridge Checkpoint Science 3 covers the requirements of your examinations in a way that I hope will help you build up your scientific knowledge and scientific enquiry skills. The questions are set to help you extract information from what you read and see, and to help you think more deeply about each chapter in the book. Some questions are set so you can discuss your ideas with others and sometimes develop a point of view on different scientific issues. This should help you in the future when new scientific issues, which are as yet unknown, affect your life.

To the teacher

Checkpoint Science 3 has been developed from *Checkpoint Biology*, *Checkpoint Chemistry* and *Checkpoint Physics* to cover the requirements of the Cambridge Checkpoint Tests and other equivalent junior secondary science courses. It also has three further aims:

- to help students become more scientifically literate by encouraging them to examine the information in the text and illustrations in order to answer questions about it in a variety of ways
- to encourage students to talk together about what they have read
- to present science as a human activity by considering the development of scientific ideas from the earliest times to the present day.

The student's book begins with a chapter called *What makes a scientist?* which looks at how the skills we develop as we grow up can be honed for studying science and for the requirements of Scientific Enquiry for stage 9 of the Cambridge Lower Secondary Science curriculum framework. This is followed by a feature on scientists past and present where the lives and works of three scientists are examined to show how they applied their scientific skills with great success.

The following chapters are arranged in sections with Chapters 1–6 addressing the learning requirements for biology stage 9, Chapters 7–12 addressing the learning requirements for chemistry stage 9 and Chapters 13–19 addressing the learning requirements for physics stage 9 of the Cambridge Lower Secondary Science curriculum framework. There is a postscript entitled *Theories of everything* which brings together areas of study from various parts of the three-year course to show how they can be linked to explain our world and how scientists throughout the ages (including today) have similarly tried to find a way of totally explaining our existence. The postscript ends with one final challenge to the students to use their creativity to devise a celebratory event for completing the course.

The student's book is supported by a teacher's resource book that provides answers to all the questions in the student's book – those in the body of the chapter and those that occur as end of chapter questions. Each chapter is supported by a chapter in the teacher's resource book which features a summary, chapter notes providing additional information and suggestions, a curriculum framework reference table, practical activities (some of which can be used for assessing Scientific Enquiry skills), homework activities, a 'lesson ideas' section integrating practical activities and homework activities, and two Practice Tests with marking guidance. A complete glossary of scientific words can be found at the back of the teacher's resource book. This can be given to students to help build up a scientific vocabulary.

Peter D Riley
October 2011

Acknowledgements

The author would like to thank Ian Lodge for reading and advising on early stages of the manuscript and Phillipa Allum, Eleanor Miles, Gina Walker, Penny Nicholson, Elizabeth Barker and Laurice Suess for their editorial work on the *Cambridge Checkpoint Science* series.

The Publishers would like to thank the following for permission to reproduce copyright material:

Photo credits

p.3 *t* © NYPL/Science Source/Science Photo Library, *b* © Antrey – Fotolia.com; **p.4** *t* © Mast Irham/epa/Corbis, *b* © David Parker/Science Photo Library; **p.5** © Sheila Terry/Science Photo Library; **p.6** Courtesy of Women in Science website; **p.7** © Neville Elder/Corbis Sygma; **p.9** © M Sweet/Getty Images; **p.10** © chasingmoments – Fotolia.com; **p.12** © Brian Gadsby/Science Photo Library; **p.14** © Chris Burrows/Garden Picture Library; **p.15** © Science Museum/Science & Society Picture Library – All rights reserved; **p.17** *l, c* © Nigel Cattlin/Getty Images, *r* © Alan Buckingham/Getty Images; **p.19** © Robert Harding Picture Library Ltd/Alamy; **p.20** © Jocika – Fotolia.com; **p.21** © Paul Harcourt Davies/Science Photo Library; **p.22** © hansenn – Fotolia.com; **p.23** *both* © Eye of Science/Science Photo Library; **p.24** © Stephen Dalton/NHPA; **p.25** *t* © Stephen Dalton/NHPA, *b* © Oxford Scientific; **p.26** *tl* © Stephen Dalton/NHPA, *tr* © Garden World Images, *bl* © John Glover/Alamy, *br* © Frans Lanting/Corbis; **p.29** *t* Courtesy of Wikipedia Commons, *b* © DEA/G. Cigolini/Veneranda Biblioteca Ambrosiana/Getty Images; **p.31** © Ornitolog82 – Fotolia.com; **p.32** © FhF Greenmedia/GAP Photos; **p.34** © Peter Scoones/Science Photo Library; **p.35** © Imagestate Media; **p.36** © Art Wolfe/Science Photo Library; **p.37** © Dr Morley Read/Science Photo Library; **p.38** © Fritz Polking; Frank Lane Picture Agency/Corbis; **p.39** © Gregory Dimijian/Science Photo Library; **p.40** © Claude Nuridsany & Marie Perennou/Science Photo Library; **p.41** © Gregory Dimijian/Science Photo Library; **p.42** © Adam Hart-Davis/Science Photo Library; **p.43** *t* © WaterFrame/Alamy, *b* © Bruce Coleman Inc./Alamy; **p.46** © Pictorial Press Ltd/Alamy; **p.49** © Hiroya Minakuchi/Getty Images; **p.50** © Dale Boyer/Science Photo Library; **p.51** *t* Biosphoto/Rauch Michel/BIOSphoto/Still Pictures, *b* © W. Perry Conway/Corbis; **p.52** © JackJelly/iStockphoto; **p.53** © Tom McHugh/Science Photo Library; **p.56** © Premaphotos/Alamy; **p.60** © Alain Compost/Photolibrary; **p.61** © National Geographic/Getty Images; **p.62** © Tui De Roy/Getty Images; **p.66** © Mary Evans Picture Library; **p.67** *r* © Giles Angel/Natural Visions, *l* © Imagestate Media; **p.70** © Getty Images; **p.71** © Anne Katrin Purkiss/Rex Features; **p.73** © GeoScience Features Picture Library/S.Bogdanov; **p.74** © Ecoscene/Alexandra Jones; **p.75** © James Balog/Aurora Photos/Corbis; **p.77** © Doug Martin/Science Photo Library; **p.79** © dzain – Fotolia.com; **p.80** © Photoshot Holdings Ltd/Alamy; **p.81** © pictureguy32 – Fotolia.com; **p.82** © alessandrozocc – Fotolia.com; **p.84** © Max Tactic – Fotolia.com; **p.87** *t* © Dr. Gopal Murti/Getty Images, *b* © Dolnikov – Fotolia.com; **p.88** © Andrew J. Martinez/Science Photo Library; **p.89** © Gerry Ellis/Minden Pictures/FLPA; **p.94** © James King-Holmes/Science Photo Library; **p.99** *l* © Science Source/Science Photo Library, *r* © A. Barrington Brown/Science Photo Library; **p.100** *l* © Jim Sugar/Corbis, *r* © Ron Frehm/AP/Press Association Images; **p.101** *l* © Wildlife GmbH/Alamy, *r* © Visuals Unlimited, Inc./Nigel Cattlin/Getty Images; **p.104** © M.I. Walker/Science Photo Library; **p.105** © P.H. Plailly/Eurelios/Science Photo Library; **p.106** © Sheila Terry/Science Photo Library; **p.107** © Sheila Terry/Science Photo Library; **p.108** © Kevin Schafer/Alamy; **p.110** © Alexandr – Fotolia.com; **p.111** © Martyn F. Chillmaid/Science Photo Library; **p.113** © Gamma-Rapho via Getty Images; **p.114** © Science Photo Library; **p.115** © North Wind Picture Archives/Alamy; **p.121** © North Wind Picture Archives/Alamy; **p.122** © Science Photo Library; **p.123** *t* © Science Photo Library, *b* © Interfoto/Alamy; **p.124** © Science Museum Library/Science & Society Picture Library – All rights reserved; **p.127** © Science Photo Library; **p.129** *l* © GeoScience Features Picture Library/Dr.B.Booth, *r* © Photodisc/Getty Images; **p.130** © Colin Keates/Getty Images; **p.131** *t* Courtesy of The Fundimentals Of Photography, C. E. K. Mees, *b* © Josh Liba/Getty Images; **p.132** © David Parker/Science Photo Library; **p.135** © fox_krol – Fotolia.com; **p.136** *t* © John Boud/Alamy, *c* © John and Lisa Merrill/Corbis; **p.138** © Kumar Sriskandan/Alamy; **p.143** © Wally McNamee/Corbis; **p.145** *l* © Lawrence Migdale/Science Photo Library, *r* © Curtis Kautzer – Fotolia.com; **pp.146–8** *all* © Andrew Lambert Photography/Science Photo Library; **p.149** © Martyn F. Chillmaid; **pp.150–1** © Andrew Lambert Photography/Science Photo Library; **p.153** © shaga – Fotolia.com; **p.154** © Andrew Lambert Photography/Science Photo Library; **p.155** © Martyn F. Chillmaid; **p.157** © Sarah Bandukwala/Fotolia.com; **p.159** *t* © Colin Molyneux/Getty Images, *b* © Bob Battersby; **p.165** © Astrid & Hanns-Frieder Michler/Science Photo Library; **p.169** © Photodisc/Getty Images; **p.170** © PhotoLink/Getty Images; **p.174** © Andrew Lambert Photography/Science Photo Library; **p.179** © Robert Harding Picture Library/Bildagentur Schuster; **p.180** © OlgaLIS – Fotolia.com; **p.182** © Robert Harding Picture Library/Bildagentur Schuster/Herbst; **p.185** *both* © Andrew Lambert Photography/Science Photo Library; **p.186** © Sheila Terry/Science Photo Library; **p.189** © TopFoto.co.uk; **p.193** © Martyn F. Chillmaid; **p.196** © Science Museum/Science & Society Picture Library – All rights reserved; **p.199** © Robert Harding Picture Library/Robert Francis; **p.200** © Science Photo Library/Peter Menzel; **p.201** © ICP-UK/Alamy; **p.206–7** © Andrew Lambert Photography/Science Photo Library; **p.208–13** *all* © Martyn F. Chillmaid; **p.214** © Martyn F. Chillmaid/Science Photo Library; **p.215** *t* © Andrew Lambert Photography/Science Photo Library, *b* © Stephan Hoerold/iStockphoto; **p.217** *both* © Andrew Lambert Photography/Science Photo Library; **p.228** © Photographic Services, Shell International Ltd; **p.229** © GeoScience Features Picture Library/D.Bayliss; **p.230** © Martin Smith/Still Pictures; **p.231** © John Cancalosi/Peter Arnold/Still Pictures; **p.232** © Harvey Lloyd/Peter Arnold/Still Pictures; **p.233** *t* © Mary Evans Picture Library, *c* © Science Photo Library; **p.234** © The Art Archive; **p.235** *c* © Martin Bond/Science Photo Library, *b* © NASA/Science Photo Library; **p.236** © Henning Bagger/epa/Corbis; **p.237** © Jim Wark/Peter Arnold/Still Pictures; **p.241** © Ocean/Corbis; **p.242** © Patrice Loiez, CERN/Science Photo Library

t = top, *b* = bottom, *l* = left, *r* = right, *c* = centre

Every effort has been made to trace all copyright holders, but if any have been inadvertently overlooked the Publishers will be pleased to make the necessary arrangements at the first opportunity.

What makes a scientist?

- The development of scientific activities
- Scientific enquiry – considering ideas and evidence
- Scientific enquiry – planning
- Scientific enquiry – obtaining and presenting evidence
- Scientific enquiry – considering evidence and approach
- Scientists past and present

A scientist grows up

You have been studying science in secondary school for two years now, so when do you think people become scientists? Figure 1 on page 2 shows a time line to help you decide.

The signs of a scientist

Babies wouldn't look and listen if they were not curious about their surroundings so we may say that curiosity is one of the first signs of being a scientist.

Later as children grow and learn to speak they begin to ask questions about what they see and hear. They want to know how things work – like machines, for example – and why events such as night and day happen.

As children grow, they begin to wonder if they could perhaps make something work better and begin to think creatively about how they could improve it. Later when they begin making scientific investigations at school they use their imagination and creativity to help them plan their work.

When investigations become more complicated, as they take longer to carry out and more data needs to be collected, the investigators may develop patience and perseverance. For example, should apparatus such as a stop watch fail to work properly they will replace it and patiently begin again.

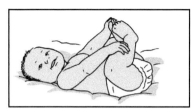

Some people believe that everyone behaves like a scientist as soon as they are born. When newborn babies are awake they watch and listen to the world around them. They makes observations.

As babies grow, they pick up things and watch them fall. They perform a simple kind of experiment. Later they bang objects and listen to their sounds. At this time many activities are investigations.

When children learn more about the world at school they are struck with a sense of awe and wonder. This makes them to want to find out more.

At secondary school, the activities in science laboratories can take the sense of awe and wonder to new heights, as magnesium burns brightly or microscopic organisms swim into view under a microscope, for example.

After learning about biology, chemistry and physics in detail at school, some people develop a great interest in one area of science and want to find out more. They continue their studies at a college or university and may then go on to work as scientists.

Figure 1 We are all scientists, from a very early age.

For discussion

At what time in their lives do you think people become scientists? Explain your answer.

Which signs of a scientist do you have? Give examples to support your answer.

The combination of curiosity, imagination, creativity and perseverance can lead to a person wanting to spend their lives learning more about scientific subjects that they have become greatly interested in.

1 You see a tree fall down in a high wind and hear it crash to the ground. If the tree was a long way away, and nobody could see or hear it, would it still make a sound as it hit the ground? Explain your answer.
2 How could you test your answer to question **1**?
3 Name two signs of a scientist that you displayed when you answered question **1**.

Scientific enquiry

Scientific enquiry is the key activity in which scientists take part to make discoveries.
A scientific enquiry is divided into four stages and in each stage there are three or more
scientific activities, as shown here.

Stage 1: Considering ideas and evidence

- Discuss and explain the importance of
  questions, evidence and explanations using
  historical and contemporary examples.
- Test explanations by using them to make
  predictions and then evaluating these against
  evidence.
- Discuss the ways that scientists work today
  and how they worked in the past, including
  reference to experimentation, evidence and
  creative thought.

Figure 2 Scientific enquiry began 1200 years
ago when Muslim scholars began testing their
ideas with investigations.

Stage 2: Planning

- Select ideas and produce plans for
  testing based on previous knowledge,
  understanding and research.
- Suggest and use preliminary work to
  decide how to carry out an investigation.
- Decide whether to use evidence from
  first-hand experience or secondary
  sources.
- Decide which measurements and
  observations are necessary and what
  equipment to use.
- Decide which apparatus to use and
  assess any hazards in the laboratory, field
  or workplace.
- Use appropriate sampling techniques
  where required.

Figure 3 Scientists choose and assemble apparatus once they
have decided what evidence and observations need to be made.

Stage 3: Obtaining and presenting evidence

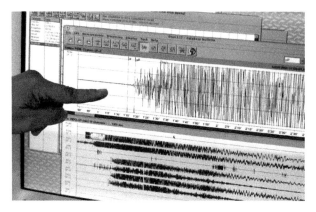

Figure 4 A seismograph is used to measure the vibrations of the Earth's surface. This one is showing vibrations due to an earthquake which measured over 9 on the Richter scale. The scale is named after Charles Richter who invented it to measure the energy released in earthquakes.

- Make sufficient observations and measurements to reduce error and make results more reliable.
- Use a range of materials and equipment and control risks.
- Make observations and measurements.
- Choose the best way to present results.

Stage 4: Considering evidence and approach

- Describe patterns (correlations) seen in results.
- Interpret results using scientific knowledge and understanding.
- Look critically at sources of secondary data.
- Draw conclusions.
- Evaluate the methods used and use this to refine methods for further investigations.
- Compare results and methods used by others.
- Present conclusions and evaluation of working methods in different ways.
- Explain results using scientific knowledge and understanding, and communicate this clearly to others.

Figure 5 Studying the data in charts such as those shown here led Alec Jeffreys to develop DNA fingerprinting – a technique that is now widely used in forensic science to help solve crimes throughout the world.

To help you become familiar with these types of activities, there are Scientific Enquiry Spotter questions in every chapter of the book. You can identify them with the icon shown below and by the green background to the question boxes. When you find one, turn back to these pages and use the activity lists to help you answer.

Scientists past and present

Our scientific knowledge has been built up by the work of scientists over the last few hundred years and by scientists working today. In this section we shall look at some of the work of a scientist from the distant past, one from the recent past and one who is living today and see what signs of a scientist we can see in their lives and what features of scientific enquiry they used in their work.

Alhazen's first investigations on light

Alhazen (also known as Abu Ali al-Hasan ibn al-Hasan ibn al-Haytham) was a Persian scientist who lived about 1000 years ago. He read about the ideas of the Ancient Greeks and became interested in many subjects such as astronomy, physics, mathematics, medicine and engineering.

He discovered that the Ancient Greeks believed that we see objects around us because light rays leave the eye and hit them, making them visible. Alhazen thought about all the objects that we can see and then considered the stars. When he looked up at the sky, shut his eyes and then opened them, he could see the stars immediately. He knew that the stars were a long way away and believed they were just too distant for rays from the eyes to reach them in an instant.

Alhazen believed that the reverse of the Ancient Greeks' idea was true – that light comes from the objects we see. He also believed that rays of light travel in straight lines. He set about testing his idea by setting up a dark room with a small hole in one wall. In a dark area outside the room he hung two lanterns, and then went back inside. He saw two spots of light on the opposite wall to the hole. He reasoned that each spot came from one of the lanterns and when he checked their positions he found that each lantern, hole and spot were in a straight line.

He then covered each lantern in turn with a cloth and discovered that the spot of light it produced became darker but when he removed the cover the spot became bright again. From this he reasoned that light does not come from the eye but from objects around it that produce light.

The evidence provided by Alhazen's work was later used by scientists in Europe to build up our knowledge about light.

Figure A Alhazen was particularly interested in investigating how light travels.

1 From looking at Alhazen's interests, would you say that he was curious? Explain your answer.
2 In Alhazen's investigation, do you think the cloth with which he covered the lanterns was opaque or translucent? Explain your answer.

3 What evidence did Alhazen first consider when thinking about light?
4 What creative thought did he apply to the evidence?
5 What do you think Alhazen decided he needed in order to test his ideas?
6 Why do you think he used two lanterns?

Anna Mani and the study of the weather

Anna Mani was born in India in 1918. She loved reading and on her eighth birthday she asked for a set of encyclopaedias for a present. By the time she was 12 she had read all the books in her local library. Anna thought about becoming a doctor but found that she particularly enjoyed physics and went to college in Madras to study it. In one set of investigations, she studied how diamonds and rubies absorb light and fluoresce. This involved exposing photographic film to the gemstones for up to 20 hours and during this time Mani did not leave her experiments but slept in the laboratory with them!

Mani's work won her a scholarship, which allowed her to travel to England and continue her study of physics at Imperial College, London. This time she studied how to improve the accuracy of meteorological instruments such as thermometers, rain gauges and anemometers. After this Mani returned to India and brought together scientists and engineers to make both instruments she had studied before and instruments that record their own data such as the thermograph.

Figure B Anna Mani was best known for her work as a meteorologist.

From her studies on the weather, Mani became interested in the idea of using energy in sunlight and the wind as a source of power to generate electricity. In 1957–1958 she set up stations around India to measure how solar radiation varied with the seasons. This data was used to plan ways to capture energy from sunlight. Mani also organised the collection of data on the wind from over 700 weather stations across India, which has been used to plan the setting up of wind farms. Around 1960, her weather studies also led her to become interested in the ozone layer and she devised equipment to measure the amount of ozone in the atmosphere.

In addition to her studies, Anna Mani enjoyed the company of other scientists and the development of science in her country and around the world. She was a member of the Indian National Science Academy, the International Solar Energy Society and the American Meteorological Society. Mani, like most scientists, had other interests outside science, which helped her relax, and she particularly liked trekking and bird-watching. Anna Mani died in 2001.

7 Did Anna Mani show signs of being curious when she was young? Explain your answer.

8 Patience and perseverance show dedication to an activity. When did Mani show great dedication to her work?

9 When was Mani's work mainly involved with obtaining and presenting evidence?

10 When did Mani's work involve creativity on a large scale?

11 What sources of secondary data did Mani's work provide?

Trevor Bayliss and the clockwork radio

Trevor Bayliss was born in London, England, in 1937 and became a swimmer, taking part in international competitions by the age of 15. He later got a job in a laboratory and also studied engineering at a local college. He went on to apply his scientific skills in the development of swimming pools. He combined his love of swimming and his work in swimming pool development by providing entertainment through swimming and work as a stunt man in a glass-sided pool. He made many friends among stunt performers and when some of them became injured and no longer able to work he began inventing devices to help them, called Orange Aids.

In 1989, Bayliss watched a television programme about AIDS in Africa. The programme stated that it was believed that the spread of the disease could be slowed if people could be educated about it by means of radio programmes. The problem was that radios needed either batteries or a power supply – and millions of people at risk of getting AIDS were too poor to afford batteries or lived in villages in the countryside that did not have a power supply. Bayliss immediately thought about a way to solve the problem and set about inventing a radio that did not need an outside source of electricity.

At first, he connected a radio to an electric motor and then connected a handle to the electric motor. When an electric motor receives a supply of electricity it turns but if the motor is turned with a handle it generates electricity. Bayliss checked that when he turned the handle the motor would make electricity and the electrical energy would make the radio work.

The problem was that he had to keep turning the motor to make the radio work. He then thought of a way to store the energy from turning the handle so he could listen to the radio without having to make it work. He used a clockwork spring. It was wound up by turning the handle and then allowed to unwind, and as it did so its stored energy passed to an electrical generator, which then supplied electricity to make the radio work. The first version of this clockwork radio could play for 14 minutes after the spring had been wound up for two minutes.

Since then, wind-up radios have been developed that can charge up rechargeable batteries and some also have solar cells.

Figure C Trevor Bayliss invented a radio that needs no batteries or mains electricity.

12 How did Trevor Bayliss show that he was curious about different things in his early life?

13 Name two things in Bayliss's life, not connected with science, that eventually led to him caring for the disabled.

14 What stimulated Bayliss into devising the clockwork radio?

15 What question might 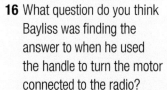 Bayliss have asked himself when thinking about his new radio?

16 What question do you think Bayliss was finding the answer to when he used the handle to turn the motor connected to the radio?

17 What question do you think Bayliss was finding the answer to when he used the clockwork spring?

For discussion

Do you think it is a good idea for scientists to have other interests besides science, as Anna Mani did? Explain your answer.

Do you think it is a good idea for scientists to have interest in science subjects other than the one they are working in, like Alhazen? Explain your answer.

◆ SUMMARY ◆

◆ Curiosity, imagination, creativity and perseverance are important characteristics of a scientist (*see page 2*).

◆ Scientific enquiry involves four stages: Considering ideas and evidence, Planning, Obtaining and presenting evidence, and Considering evidence and approach (*see page 3*).

◆ Over many centuries, scientists throughout the world have worked to broaden and deepen our scientific knowledge and understanding, by applying the four stages of scientific enquiry, and they continue to do so (*see pages 5–7*).

End of chapter questions

A wind turbine can start to generate electricity when the wind speed reaches 16 km/h and generates most electricity at 48 km/h but must be stopped once the wind reaches 104 km/h as this speed can cause serious damage.

Here is a table showing the wind speeds in four seasons of the year at three different places – A, B and C.

Place	Average wind speed in each season / km/h			
	spring	summer	autumn	winter
A	85	52	44	60
B	50	40	10	40
C	60	45	78	120

1 Which place is the windiest?

2 Which is the least windy place?

3 Which place and season has a wind speed at which the wind turbine would generate the most electricity?

4 In which place and season would the generator have to be stopped?

5 Which place would be best for maximum electricity generation in summer? Explain your answer.

6 Which place would be best for generating electricity all year round? Explain how you made your decision.

BIOLOGY

1 Photosynthesis

◆ The presence of starch in a leaf
◆ Water and plant life
◆ Carbon dioxide and starch production
◆ Light and starch production
◆ Chlorophyll and starch production
◆ Photosynthesis and oxygen production
◆ The word equation for photosynthesis
◆ Mineral salts and plant growth

Photosynthesis is the process by which plants make carbohydrate as food. The carbohydrate is stored in the form of starch. In order to investigate photosynthesis, we have to use a number of chemicals, a range of apparatus and the form of radiation energy known as light. This means that although photosynthesis is a biological process we have to use some knowledge of chemistry and physics in order to understand it.

Figure 1.1 Photosynthesis is occurring in the leaves of these palm trees in Jamaica as the Sun shines down on them.

Starch in leaves

Starch is a store of energy. It is a **carbohydrate** and forms colourless grains in the cytoplasm of leaf cells in the palisade and spongy mesophyll layers.

The test for starch

The presence of starch grains in a leaf can be revealed in the following way.

1 A beaker of water is boiled and the heat source is turned off.
2 The leaf is held in a pair of forceps and dipped in the beaker of hot water for about 30 seconds. This kills the cells and prevents any further reactions taking place in them. It also makes it easier for iodine solution to enter the cells in step 5.
3 The leaf is then placed in the bottom of a test tube and covered with ethanol. The test tube is placed in the beaker of hot water as Figure 1.3 shows. Alcohol has a boiling point that is below that of the hot water and as it boils it dissolves most of the chlorophyll in the cells and makes the starch grains easier to see when they are stained with iodine.

1 Draw one of the palisade cells in Figure 1.2 and label five of its parts.
2 In which food category is starch? What elements is it made from?

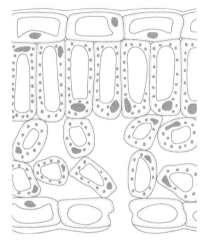

Figure 1.2 The layers of cells in a leaf

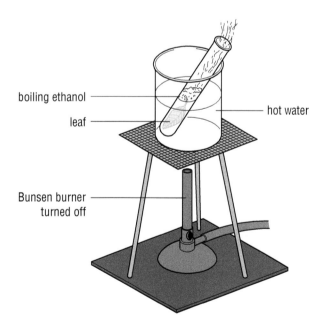

Figure 1.3 Removing chlorophyll from a leaf

boiling ethanol
leaf
hot water
Bunsen burner turned off

3 Plant cells build up starch and then remove it, as we shall see later. When testing a leaf for starch, what might happen if the cells were not killed in step 2?
4 What would happen if the boiling point of ethanol was above that of water?
5 What colour would you expect the ethanol to go after the leaf has been boiled in it? Explain your answer.
6 What might happen to the brittle leaf if you tried to spread it out on the white tile?

4 The alcohol in the test tube is poured into a second beaker and the leaf, which is now brittle, is removed with forceps and dipped into the hot water again to make it softer and easier to handle in the next step.

7 Why is it important to spread the leaf out on a white tile and not a coloured one?

8 Why is it important that a plant should be destarched before it is used in photosynthesis investigations?

5 The soft leaf is then spread out on a white tile and drops of iodine are released onto its surface. The iodine solution enters the cells and if starch grains are present they are stained blue-black.

A destarched plant

A plant from which the starch has been removed is called a destarched plant. A plant is destarched by putting it in a dark place, such as a cupboard, for two or three days. Once a plant is destarched it is ready to take part in investigations about photosynthesis.

Constructing the equation for photosynthesis

Water

Water is essential for life. If plants do not have water, they die.

Figure 1.4 These maize plants in South-West France died through lack of water in a drought.

Water is drawn into the plant roots from the soil and is carried along microscopic tubes in the xylem tissue from the root, into the stem and then to all the other parts of the plant. When it reaches the leaves it takes part in the making of food.

The first step in building up the equation for photosynthesis begins with water, as shown here:

water +

Carbon dioxide and starch production

To investigate the role of carbon dioxide in starch production, two destarched plants are needed. One is enclosed in a transparent plastic bag with a dish of soda lime and the other is enclosed in a transparent plastic bag with a dish of sodium hydrogencarbonate solution. The air in the bag is sealed in with an elastic band holding the bag to the plant pot (Figure 1.5).

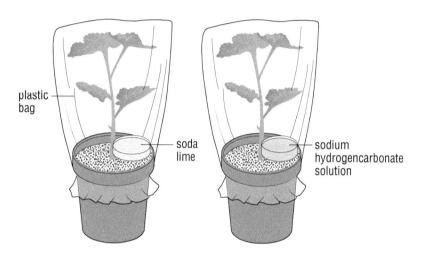

Figure 1.5 Investigating the effect of carbon dioxide on starch production

Soda lime is a mixture of sodium hydroxide and calcium hydroxide. It has the property of absorbing carbon dioxide from the air. Sodium hydrogencarbonate solution releases carbon dioxide into the air. It enriches the air with carbon dioxide.

The plants are set up in a sunny place and left for a few hours and then a leaf from each one is tested for starch. A leaf from the plant in the air without carbon dioxide does not contain any starch but a leaf from a plant in the air enriched with carbon dioxide does contain starch. So we can conclude that carbon dioxide is needed for starch production.

The second step in building the equation for photosynthesis is shown here:

$$\text{water} + \text{carbon dioxide} \rightarrow \text{carbohydrate (starch)}$$

So far we have produced a word equation for photosynthesis with two reactants and one product but before we investigate the final product we need to look at two other factors involved in photosynthesis and see if we can fit them into the equation – light and chlorophyll.

9 If you were to test a small sample of soda lime with universal indicator would it register a pH above or below 7? Explain your answer.

10 What is the purpose of the plastic bag and elastic band?

11 Is it important for the bags to be transparent? Explain your answer.

Light and starch production

One destarched plant is needed for this investigation to find out if light is needed for starch production. One leaf is covered with aluminium foil, which is an opaque material. Another leaf is covered in transparent plastic (Figure 1.6)

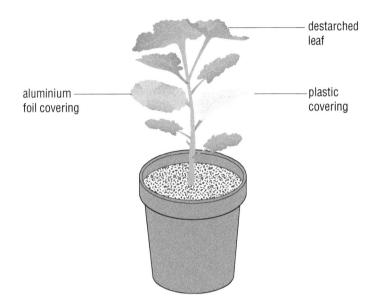

Figure 1.6 Investigating the effect of light on starch production

12 Why do you think a cover was used for the leaf exposed to the light?

The plant is left for four hours in a sunny place. Then the covers are removed from the leaves and they are tested for the presence of starch. The leaf enclosed in aluminium foil does not contain any starch, while the leaf enclosed in transparent plastic does contain starch. This shows that light is needed for starch production and the next step in building the equation for photosynthesis is:

$$\text{water} + \text{carbon dioxide} \xrightarrow{\text{light}} \text{carbohydrate}$$

Chlorophyll and starch production

A plant with white patches on its leaves is needed to investigate the role of chlorophyll in starch production. Plants with green and white leaves are known as variegated plants.

The variegated plant is destarched and then placed in the light for over four hours. When a leaf is removed and tested with iodine solution, the green parts where chlorophyll is present turn blue-black but the white parts, which do not contain any chlorophyll, do not turn blue-black.

Figure 1.7 This plant has variegated leaves. The white parts show where chlorophyll is absent.

This shows that chlorophyll is needed for starch production and the next step in constructing the equation is:

$$\text{water} + \text{carbon dioxide} \xrightarrow[\text{chlorophyll}]{\text{light}} \text{carbohydrate}$$

Photosynthesis and oxygen

Joseph Priestley (1733–1804) was an English chemist who studied gases. The apparatus used at the time to trap gases was an upside-down container put over the experiment to catch any gases that were produced. In the course of his investigations, Priestley discovered that sometimes a gas was produced in which things could not burn. When he put a plant such as mint in a jar of this gas and let sunlight shine on it, he found that the gas appeared to change to one that did allow things to burn in it. Priestley considered that a gas had been produced which refreshed the air.

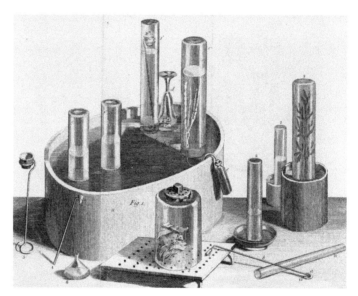

Figure 1.8 This is the apparatus Joseph Priestley used in his investigations on gases.

Later Priestley met the French chemist Antoine Lavoisier (1743–1794) and told him about his discovery. After thinking about Priestley's investigations and performing some tests on the gas, Lavoisier named it oxygen.

Today the gas produced by plants can be investigated using Canadian pondweed set up as shown in Figure 1.9. The apparatus on the left is placed in a sunny place and the other apparatus is kept in the dark. After about one week the amount of gas collected in each test tube is examined. The plants in the dark have not produced any gas but the plants in the light have produced a gas that relights a glowing splint – showing that the gas is oxygen.

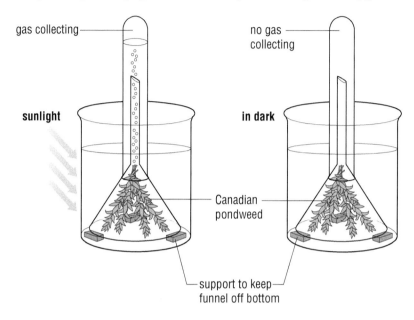

13 Compare the apparatus used by Priestley with the set-up used today to demonstrate oxygen production in plants.

14 Word equations have reactants and products. What are the reactants and products in the word equation for photosynthesis?

Figure 1.9 Apparatus for investigating oxygen production

This information provides the final step in producing the word equation for photosynthesis:

$$\text{water + carbon dioxide} \xrightarrow[\text{chlorophyll}]{\text{light}} \text{carbohydrate + oxygen}$$

Mineral salts

When chemists began studying plants, they discovered that they contain a wide range of elements. With the exception of carbon, hydrogen and oxygen, plants obtain these elements from mineral salts in the soil. Three examples of minerals in the soil are nitrogen in the form of nitrate salts, phosphorus in the form of phosphate salts and potassium in the form of salts such as potassium nitrate.

Scientists found out about the importance of each mineral in the following way. They made a solution of salts with all the minerals that plants need, except the one they were investigating. For example, if the effect of nitrogen was being investigated, they made a solution that contained all the minerals except nitrogen. The plant was then grown for a few weeks in this solution (Figure 1.10), alongside a plant growing in a solution containing all the mineral salts required, and the differences between the two plants were recorded. From these studies, scientists discovered the following.

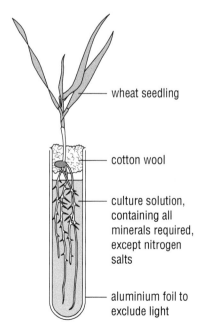

Figure 1.10 This plant has been set up in a solution from which nitrogen salts have been omitted.

15 Put the information about mineral salts and their uses by plants into a table. Include information about what happens if the mineral salt is missing.

16 What mineral might be missing if the leaves go yellow?

17 Why might a plant show poor growth?

18 When investigating the importance of different mineral salts, why was a solution used for each experiment, rather than a mixture of soil and the solution?

- Nitrogen is needed for the development of the leaves. Without nitrogen, the leaves turn yellow and the plant shows poor growth. Further studies have shown that nitrogen is needed for making the green pigment chlorophyll and for making proteins that form part of the structure of the plant.
- Phosphorus is needed for the development of the roots. Without phosphorus, a plant shows poor growth. Further studies have shown that phosphates are needed to help plants make food by photosynthesis and to respire.
- Potassium is needed for the development of flowers and fruits. If potassium is absent the leaves become yellow and grow abnormally. Further studies have shown that potassium helps the plant make chlorophyll and proteins forming part of the structure of the plant.

lack of nitrogen

lack of phosphorus

lack of potassium

Figure 1.11 Plants showing mineral deficiency

The path of minerals through living things

When animals eat plants, they take in the minerals in the plant tissues and use them in their bodies. Some of the minerals are released in the solid and liquid wastes (dung and urine) that animals produce. As bacteria feed on these wastes, the mineral salts are released back into the soil. The mineral salts are also released when the plants and animals die and decomposers break down their bodies. Plants, animals and their wastes are **biodegradable**. This means they can be broken down into simple substances that can be used again to make new living organisms. These simple substances have been recycled since the beginning of life on Earth.

19 Could you be a recycled dinosaur? Explain your answer.

◆ SUMMARY ◆

◆ The presence of starch in a leaf can be identified in a test using iodine solution *(see page 11)*.

◆ A plant is destarched by being kept in the dark *(see page 12)*.

◆ Plants need water for photosynthesis *(see page 12)*.

◆ Carbon dioxide is needed for starch production *(see page 13)*.

◆ Light is needed for starch production *(see page 14)*.

◆ Chlorophyll is needed for starch production *(see page 14)*.

◆ When plants photosynthesise they produce oxygen *(see page 15)*.

◆ A word equation can be written to summarise the process of photosynthesis *(see page 16)*.

◆ A range of mineral salts are needed for healthy plant growth *(see page 16)*.

End of chapter question

How could you demonstrate to farmers that using dung on the soil can improve their crops?

2 Reproduction in flowering plants

◆ The reproductive parts of a plant
◆ Insect- and wind-pollinated flowers
◆ Pollen grains and pollination
◆ Fertilisation
◆ Seed formation
◆ Seed and fruit dispersal

For discussion

Which plants are in flower where you live? How many people in your class can name six local plants? What is the most that anyone can name?

What was the last plant that you really looked at? Probably, like most people, you cannot remember and yet the chances are there are many plants in your surroundings – in the school grounds, on your journey from home to school and perhaps around your home and in it too. Plants are such a large part of our world that we tend to ignore them but when they come into flower we cannot fail to notice them.

Figure 2.1 Plants are often grown for the colour of their flowers.

The parts of a flower

The flower is the reproductive structure of the plant. It contains the reproductive organs. Many plants have flowers, which contain both the male and female reproductive organs. During reproduction, pollen (page 22) passes from

19

Figure 2.2 Petals forming a circle. The structure of this type of flower is shown in Figure 2.3.

flower to flower in a process called **pollination**. There are two main types of flower – insect-pollinated flowers and wind-pollinated flowers.

Insect-pollinated flowers

The form of insect-pollinated flowers can vary enormously but two common types are flowers which have **petals** that open in a circle and flowers in which the petals form a tube. Figures 2.2 and 2.3 show flowers whose petals open to form a circle while Figures 2.4 and 2.5 show the parts of a flower with a tubular form.

The **sepal** is a small, tough, leaf-like structure. When the flower is in bud, a group of sepals lay over each other, like tiles on a roof, to protect it. The sepals form a ring called the **calyx**.

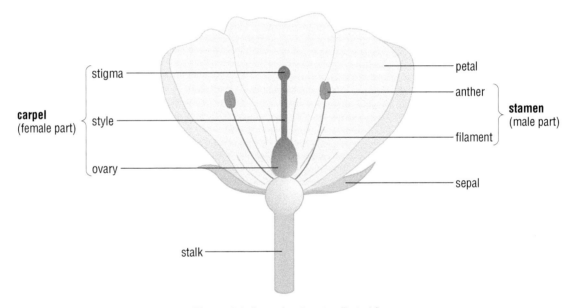

Figure 2.3 Parts of an insect-pollinated flower

Petals are found in plants that use insects to pollinate their flowers. They are a large and colourful part of the flower. Some petals produce scent, and the colour and the scent attract insects to the flower.

The petals form a ring called the **corolla**. The petals in the corolla may form a circle or a tube. Inside the corolla are the **stamens**. They are the male part of the flower. Each stamen has two parts – the stalk, called the **filament**, and the pollen-producing organ, called the **anther**. Inside the ring of stamens is the female part of the flower, which is made up from one or more **carpels**. Each carpel has a pollen-receiving surface called a **stigma**. Beneath the stigma is the **style**. It is connected to the **ovary**, which contains one or more **ovules**.

1 Imagine that you are watching a flower bud burst open. In which order will you see the parts of the flower?

Figures 2.4 and 2.5 show flowers whose petals form a tube.

Figure 2.4 Petals forming a tube. The structure of this type of flower is shown in Figure 2.5.

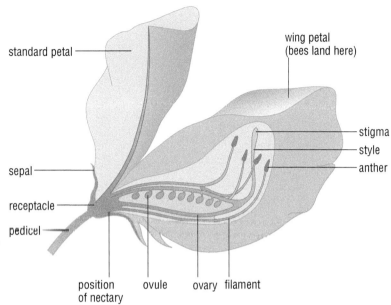

Figure 2.5 Parts of a tubular insect-pollinated flower

2 How can you tell the stigma from the stamens in the flower in Figure 2.5?

Insect-pollinated flowers have further adaptations to help them attract insects and use them for pollination. At the base of the petals there may be **nectaries**, producing a sugary liquid called nectar on which the insects feed. Many flowers produce more pollen than is needed for pollination and this may also be taken as food by the pollinating insects.

Short filaments keep the anthers inside the flower so that the insect can brush past them. The anthers of insect-pollinated flowers make a smaller amount of pollen than those of wind-pollinated flowers. The stigma is often flat and held on a short style inside the flower so that the insect can easily land on it.

Wind-pollinated flowers

Wind-pollinated flowers do not stand out like insect-pollinated flowers because they do not have colourful petals. The most common plants with wind-pollinated flowers are the grass plants.

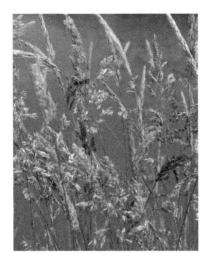

Figure 2.6 Grass plants have many small flowers grouped together at the top of a flower stalk.

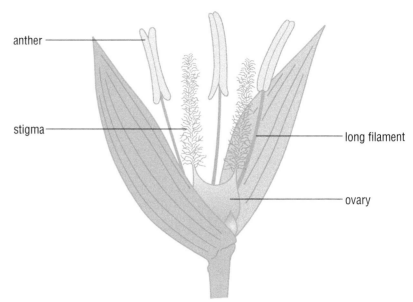

Figure 2.7 Wind-pollinated flower

3 Make a table to compare wind- and insect-pollinated flowers.

4 Why do wind-pollinated flowers produce much more pollen than insect-pollinated flowers?

Wind-pollinated flowers are smaller than insect-pollinated flowers. They may have green petals and do not produce nectar or scent. The flowers have long filaments that allow the anthers to sway outside the flower in the air currents. The anthers make a large amount of pollen and the stigma is a feathery structure that hangs outside the flower and forms a large surface area for catching pollen in the air.

Pollen grains and pollination

The sex cells of living things are called **gametes**. In animals, the male gamete is the sperm and the female gamete is the egg or ovum. In flowering plants, the male gametes are produced in the anthers where each one is enclosed in a **pollen grain**. The female gamete, called the egg cell, is in the ovule.

When the pollen grains are fully formed, the anther splits open to release them. Pollination occurs when pollen is transferred from an anther to a stigma. If the pollen goes from an anther to the stigma of the same flower or other flowers on the same plant the process is called **self-pollination**. **Cross-pollination** occurs if the pollen goes from an anther to the stigma of a flower on another plant of the same species (Figure 2.8).

5 What is the difference between self-pollination and cross-pollination?

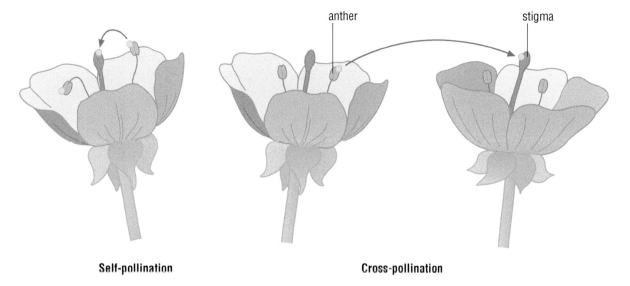

Self-pollination   **Cross-pollination**

Figure 2.8 Types of pollination in flowering plants

Most plants produce flowers that have both male and female reproductive parts. They avoid self-pollination in two ways. One way is by releasing pollen from the anthers before the stigmas of that plant are ready to receive it. Another way is by having the stigmas of the plant ready to receive pollen from other plants of the same species before its own anthers are ready to release their pollen.

As we have seen, the two main ways in which the pollen grains get from one flower to another for cross-pollination are by hitching a ride on an insect (Figure 2.2) or by being blown by the wind (Figure 2.6). Pollen grains carried by insects may have a spiky surface that helps them stick to the hairs on the insect's body. Pollen grains carried by the wind are very small and light so that they can travel easily on air currents (Figure 2.9).

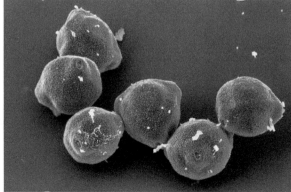

Figure 2.9 Highly magnified views of pollen from an insect-pollinated plant (left) and from a wind-pollinated plant (right)

Bees and flowers

Karl von Frisch (1886–1982) was an Austrian zoologist who studied how bees communicate with each other. In 1973 he received a Nobel Prize for his work on animal behaviour.

The scent from a flower producing nectar travels through the air. It may stimulate the receptor cells of a honey bee and the insect flies towards it. As it gets closer, the bee also uses its eyes to find the flower. Its eyes are sensitive to ultraviolet light. This makes some of the pale markings we see in normal light stand out more distinctly and helps the bee to identify the flower. Some of the markings are lines running down the inside of the petal. They are called honey guides and direct the bee towards the nectar.

After landing on the flower, the bee sticks its head between the stamens and probes the nectary with its mouthparts. While taking up the nectar, it brushes past the anthers and pollen collects on the hairs of its back. When the bee has collected the nectar it flies on to the next flower and feeds again. Some of the pollen on its back passes onto the stigma of the next flower.

The bee has stiff hairs on its front legs. Periodically it runs them through its body hair like a comb. This action collects the pollen off the bee's back and it is stored in structures called pollen baskets, which are made from hairs on its back legs.

Figure 2.10 A bee in flight showing full pollen baskets. When the bee swallows the nectar it collects in a cavity called the honey sac.

6 What attracts the bee to the flower? Which sense organs does it use?

7 How do you think that Karl von Frisch gathered information about the honey bee's behaviour?

8 How does the behaviour of the dancing bee help a colony of plants that have come into flower?

9 How do you think a hive of bees survives the winter when there are no flowers to feed on?

10 Why are hives of bees often kept in orchards?

For discussion

How useful are bees? Should we worry if there were fewer bees? Explain your conclusions.

The action of enzymes and the addition of other substances change the nectar into honey. After the bee has returned to its hive, it regurgitates the honey and passes it on to other bees working in the hive. They store it in the honeycomb. Also, the pollen is removed from the pollen baskets and stored.

The bee indicates the source of the nectar to the other bees in the hive by performing a dance on the honeycomb. The dance involves the bee moving in circles, waggling its abdomen and moving straight up and down on the vertical surface of the honeycomb. From this performance, the other bees can tell the distance, direction and amount of nectar available and can set out to search for it.

Unusual methods of pollination

Most plants either have flowers that are pollinated by insects during the day or have wind-pollinated flowers, but there are unusual methods of pollination found in many different parts of the world. Here are just a few examples.

Figure 2.11 The bird of paradise flower has stamens that form perches for birds. They pick up the pollen on their feet.

Figure 2.12 The flowers of the century plant contain large amounts of nectar for bats to drink. Each bat carries away the pollen on the fur on its head and neck.

Figure 2.13 Honeysuckle flowers produce nectar at night to attract moths to pollinate them.

Figure 2.14 Canadian pondweed produces male and female flowers on long stalks, which let them reach the water surface. The male flower releases pollen onto the water and it floats away. Some of this pollen reaches the female flowers and pollination occurs.

Figure 2.15 The cuckoo pint has male and female flowers on a short stem, covered by a hood. The plant releases an unpleasant smell to attract small flies, which enter the hood, pushing through the downward-pointing hairs. As they pass over the female flowers, the stigmas collect pollen from the insects. The hairs prevent the insects from leaving for a few days. During this time the male flowers make pollen. When they release it, the hairs wither so that the insects can crawl over the male flowers, pick up the pollen and escape to take it to other cuckoo pint plants.

Figure 2.16 *Rafflesia* is a parasitic plant that lives on the roots of vines in the Malaysian rainforests. It produces the largest known flower of any plant, 91 cm across. It is sometimes called the stinking corpse lily because of the smell of rotting meat that it produces to attract flies. There are separate male and female flowers and the flies they attract bring about pollination.

11 How could you arrange the unusual methods of pollination into groups?

12 Which flower acts as a trap?

13 Why do bat-pollinated flowers need to produce large amounts of nectar?

14 Which method carries the highest risk of failure?

Fertilisation

After a pollen grain has reached the surface of a stigma, it breaks open and forms a pollen tube. The male gamete (cell) that has travelled in the pollen grain moves down this tube. The pollen tube grows down through the stigma and the style into the ovary (Figure 2.17).

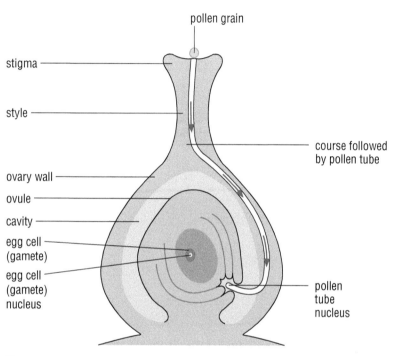

Figure 2.17 Fertilisation

15 What is the difference between pollination and fertilisation?

16 Trace the path of a male gamete nucleus from the time it forms in a pollen grain in an anther until the time it enters an ovule.

In the ovary are ovules, each containing a female gamete (cell). When the tip of a pollen tube reaches an ovule the male gamete enters the ovule. It fuses with the female gamete in a process called **fertilisation** and a cell called a **zygote** is produced.

After fertilisation

The zygote divides many times to produce a group of cells, which form a tiny plant. Structures that later become the root and shoot are developed and a food store is laid down. While these changes are taking place inside the ovule, the outer part of the ovule is forming a tough coat. When the changes are complete, the ovule has become a seed (Figure 2.18, page 28).

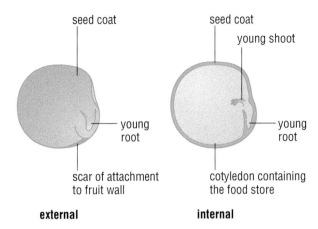

Figure 2.18 Parts of a pea seed

As the seeds are forming, other changes are taking place. The petals and stamens fall away. The sepals usually fall away too but sometimes, as in the tomato plant, they may stay in place. The stigma and style wither and the ovary changes into a **fruit**.

Early observation on plants

Theophrastus was a scholar and philosopher living in Ancient Greece between about 371 and 287 BCE. He became known as the father of botany because of the observations he made on the plant kingdom, which he recorded in two books. In these books he grouped his observations as follows.

- Plant parts
- Reproduction and sowing
- Trees
- Shrubs and spiny plants
- Herbs
- Plants that produce edible seeds
- Plants that produce useful substances such as juices, resins and gums
- Plant growth
- Factors that affect the number of new plants produced
- The properties of plants, such as their smell and how they taste

Theophrastus had a botanical garden in which he grew many different kinds of herbs. Early botanical gardens were called physic gardens because physicians (doctors) used the herbs to treat disease. Theophrastus made observations on his plants, and also travelled around Greece making further observations. He lived in the time of Alexander the Great who established an empire from Greece and Egypt in the west, through the lands where Turkey and Iran are today, to the Himalayas and the border of India in the east. People travelled through the empire trading with each other and brought information to Theophrastus about plants that grew in other lands beyond Greece. These included the cotton plant from Africa and India, the Banyan tree (a kind of fig tree) and the pepper plant (black pepper) and cinnamon plant from India. He used this information in his books.

Discovering the sexes of plants

Prospero Alpinio (1553–1617) was an Italian doctor but was also very interested in plants. He worked in Egypt for three years, and during this time he observed that there were two kinds of date palm – one that produced fruits and one that produced a dust. He observed that the fruit-producing trees only produced fruit after dust from the dust-producing trees was blown on to them. He identified the fruit-producing trees as female plants and the dust-producing plants as male plants. On his return to Italy, he became the director of the botanical garden at Padua and continued his studies on plants.

Nehemiah Grew (1641–1712) was an English scientist who studied all parts of the plant. In his work on flowers, he predicted that the stamens were the male organs of reproduction and studied the pollen of different species of plants under the microscope. From his observations he discovered that, while pollen was generally globe shaped, different species produced pollen grains with different shapes and these could be used to identify the plant.

Rudolph Camerarius (1665–1721) was a German doctor who also studied plants. From his observations and experiments, he identified the stamen as the male organ of reproduction and the ovary as the female organ of reproduction, and described how pollen produced by the stamen fertilised the ovary.

Classification using plant reproduction

Carl Linnaeus (1707–1778) became interested in plants from the age of eight and went on to become a lecturer at Uppsala University in Sweden. He made a plant study trip to Lapland and as he discovered new species he decided that he needed a way to put them into groups. He studied the work of Alpinio, Grew and Camerarius on plant reproduction and set up a way to classify plants based on the number of male and female parts, their lengths and their arrangement in the flower. This system was eventually discarded but it led Linnaeus to devise a way of classifying all living organisms, which is still used today.

Figure A Prospero Alpinio

1 What do you think was the dust of the male date palm?

2 From his observations, 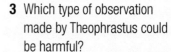 how did Theophrastus group plants according to their size?

3 Which type of observation made by Theophrastus could be harmful?

4 What primary resources did Theophrastus use when making observations?

5 What secondary sources did Theophrastus use when writing his books?

6 What instrument did Grew use in his observations that helped him study pollen?

7 Who confirmed Grew's prediction about stamens?

8 Who provided evidence for Linnaeus's work?

9 What example of creative thought did Linnaeus display in his studies of the evidence of others?

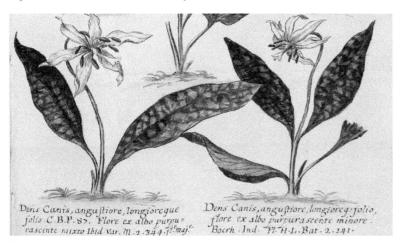

Dens Canis, angustiore, longioreque folio C.B.P. 87. Flore ex albo purpu= rascente mixto Ibid Var. M.2.344.f.maj.

Dens Canis, angustiore, longiore, folio, flore ex albo purpurascente minore. Boerh. Ind. Pl. H.L.Bat.2.141.

Figure B Plant artists made very detailed and accurate pictures of plants to help scientists classify them.

Dispersing the fruits and seeds

A plant may produce many fruits. If they were all to fall to the ground around the plant the seeds inside them could eventually grow into new plants. There would be hundreds of new plants growing close together competing with each other for light, water and minerals in the soil, and so many would die. Overcrowding is prevented by fruit and seed **dispersal**.

Plants use a range of ways to spread out their fruits and seeds so that when new plants grow they are not competing with each other. The disadvantage of dispersal is that seeds may land in unsuitable surroundings, in which they fail to grow. However, plants produce large numbers of seeds to be dispersed to increase the chance of some of them reaching suitable surroundings where they may grow into new plants.

A few plants, such as the oak tree of the northern hemisphere and the Brazil nut tree of the Amazonian rainforest, have fruits that simply drop to the ground. Some of the acorns from the oak tree are collected and stored by squirrels and mice some distance from the tree. The Brazil nut fruit has a tough outer coating which is opened by the agouti. This large rat-like mammal carries the seeds away and stores them. The squirrels, mice and agoutis return to their food stores to feed on the fruits and the seeds inside, but they forget where they have stored all of them. These 'lost' seeds may eventually germinate and grow successfully into trees without competing with other seedlings or the parent tree.

Goose grass (which grows in Asia, Europe and North America) and burdock (which grows in Asia, Europe and North Africa) have fruits with hooks on them. The hooks stick to the fur of passing mammals and they may be carried several kilometres before they are rubbed off and fall to the ground.

For discussion

How could water be used to disperse fruits of some plants? Explain your answer.

How would the fruit have to be adapted to survive?

Figure 2.19 Animal-dispersed fruits: goose grass (left) and burdock (right)

The flesh of succulent fruits often has a bright colour and is eaten by many different mammals and birds. If the seeds are small, as in berries, they are eaten with the flesh of the fruit. The seed coats are resistant to the digestive processes of the animal and the seeds leave the animal's body in the faeces. The large seed in the stone of a succulent fruit such as an apricot may also be dispersed by animals. It is not eaten but is thrown away when the animal finishes its meal.

Figure 2.20 The seeds in these berries will pass undamaged through the waxwing's body and be deposited far from the parent tree in the bird's faeces.

Many fruits with a small mass develop long hairs – for example, the willow herb, which grows worldwide. The hairs increase the air resistance of the fruit and allow it to be blown away. The dandelion, which grows in many places around the world, has a fruit that forms a tuft of hairs. This acts as a parachute and slows down the fruit's sinking speed as the wind blows it along.

Fruits with a larger mass that are dispersed by the wind have a part that is shaped into a wing. The sycamore, a type of maple tree that grows in Europe, and the South American tipu tree both have winged fruits. The large surface area of the wing catches the wind as the fruit falls from the branches and allows it to be carried away from the tree on the air currents.

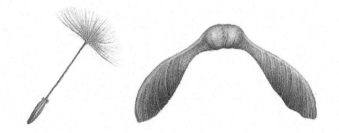

Figure 2.21 Wind-dispersed fruits: dandelion fruit (left) and sycamore fruit (right)

17 Construct a table to summarise how fruits and seeds are dispersed. Use the following terms: wind dispersal, animal dispersal, succulent fruits, hooked fruits, hard fruits, parachute fruits, winged fruits and explosive fruits. Give examples of each.

18 Why is a device that 'slows down the sinking speed' useful to wind-dispersed fruits?

19 Which kind of dispersal by animals provides the seeds with mineral salts? Explain your answer.

Some plants, such as the lupin which grows in many places around the world, and the gorse which grows in Europe and North Africa, produce fruits called pods that dry and twist. The tension in the twisting pod becomes so great that the pod splits open and shoots the seeds out.

Figure 2.22 The ripe fruit of this lupin splits suddenly, shooting the seeds away from the parent plant.

◆ SUMMARY ◆

◆ The flower contains the reproductive organs of the plant (*see page 19*).

◆ The male part of the flower is the stamen. The anther on the stamen produces pollen (*see pages 20 and 22*).

◆ The ovary is part of the female part of the flower. It contains ovules (*see page 27*).

◆ Pollination is the transfer of pollen from the anther to the stigma (*see page 22*).

◆ Wind and insect pollination are the two main kinds of pollination (*see page 23*).

◆ Fertilisation is the fusion of the male gamete nucleus with the female gamete nucleus to form a zygote (*see page 27*).

◆ The ovule develops into a seed after fertilisation (*see page 27*).

◆ The ovary develops into a fruit after fertilisation (*see page 28*).

◆ Seed dispersal reduces competition between seedlings (*see page 30*).

◆ There is a range of ways in which seeds and fruits are dispersed (*see page 30*).

End of chapter questions

Pollen grains will grow pollen tubes if they are placed in a sugar solution of a certain concentration. The results in the tables below reflect results sometimes produced in investigations.

An experiment was set up to find the concentration of sugar that would cause the pollen grains of a plant to produce pollen tubes. Here is a table of the results.

Concentration of sugar solution/%	Number of pollen grains	Number of grains with tubes
5	20	4
10	20	18
15	20	3
20	20	0

1 What percentage of pollen grains produced tubes in each solution?

2 What would you expect the concentration of sugar in the stigmas of the flowers to be?

When the pollen of a second type of plant was investigated the following results were obtained.

Concentration of sugar solution/%	Number of pollen grains	Number of grains with tubes
5	20	0
10	20	1
15	20	3
20	20	8

3 Why was it decided to take the investigation further?

4 What do you think was done to take the investigation further?

3 Adapting to a habitat

Every living thing has a certain basic body plan. For example, a flowering plant has a root, a stem, leaves and flowers, and a bird has a head with a beak and a body with two wings and two legs and a tail. However, every living thing also has special features that help it live in its surroundings. These features are called **adaptations**. For example, a flowering plant that grows in a windy habitat has a short stem, which means there is less chance of it being blown over. The short stem is an adaptation to the habitat. A heron is a bird with very long legs and a pointed, spear-shaped beak. These two features are adaptations of the heron's basic body plan. They help the heron survive by allowing it to wade into deep water and spear fish with its beak.

Figure 3.1 A heron fishing in a river

Adaptations to the seasons

There are only a few habitats, such as caves, where the environmental conditions remain the same throughout the year. In most habitats there are periods in the year called seasons during which the weather has a particular feature. A habitat might have a dry season and a wet season, or the cold weather of winter and the warm weather of summer. Below are some examples of habitats with seasonal weather.

Seasons in a European woodland

In Europe, there are four seasons of the year – winter, spring, summer and autumn.

Deciduous trees adapt to the cold, icy weather of winter by losing their leaves. Deciduous trees have large, flat leaves, which lose a great deal of water. In winter, the ground is often frozen so the water cannot pass into the roots. If the trees kept their leaves, they would lose water but would not be able to replace it. They would dry out and die.

Most insects spend winter in a stage of their life in which they do not need to move or search for food. These are the egg and pupa stages. Animals such as hedgehogs and bats, which feed on insects, hibernate during winter.

In spring as the ground warms up, woodland plants such as snowdrops and bluebells grow from bulbs and produce leaves and flowers. These plants use the sunlight shining through the bare branches to make food, and early-hatching insects to pollinate their flowers. As spring goes on the deciduous trees put out their leaves and flowers. Hibernating animals wake up and search for food. Woodland birds build nests and begin to rear young.

Figure 3.2 A European woodland in spring

1 Why is it an advantage for a woodland plant to grow before the trees come into leaf?

2 Why don't insects such as butterflies spend the winter in their active stages of caterpillar and adult?

3 Why is it an advantage for birds and other animals to rear as many young as they can in the summer?

4 State two ways in which losing leaves in the autumn helps deciduous trees.

5 What are the advantages of a large group of animals moving around together in their habitat?

6 A zebra has a height at the shoulder of 1.2–1.4 m. A wildebeest has a shoulder height of 1.0–1.3 m. A gazelle has a shoulder height of 51–89 cm. Use this information and the information in the text to describe and explain the order in which the animals would move through an area of long grass.

7 Do you think carnivores such as lions migrate too? Explain your answer.

In summer the leaves of the trees form a shady canopy over the woodland floor. Few plants are in flower there now. The birds may lay second or even third clutches of eggs and raise more young. Caterpillars feed on the leaves.

In the autumn the weather becomes cooler again. Trees produce fruits such as nuts and berries, which are eaten by many mammals and birds. The leaves of the deciduous trees lose their chlorophyll. At this time, brown and yellow pigments in the leaves give them their colour. The trees release waste products into their leaves, and in time the leaves fall. Animals that hibernate gorge themselves on food to build up fat. This is the energy store which they will use through the winter months in order to stay alive.

Seasons on an African grassland

In the Serengeti National Park in East Africa there is a wet and a dry season. The plains are covered by long grass, which is eaten by huge numbers of herbivores. Zebras eat the tough tops to the grass stalks, which contain the wind-pollinated flowers. Wildebeest (gnu) feed on the more succulent leaves lower down the plant, while the young shoots and the seeds on the ground are eaten by gazelles.

During the wet season the animals migrate to a drier part of the plains in the south. At the beginning of the dry season they move to the west where there is a little rainfall and the grass is still thick. In the middle of the dry season they move to a region where the soil is particularly fertile and the plants are still edible.

Figure 3.3 A herd of zebras migrating to a more favourable part of the plains

Adaptations to a habitat

Each habitat has a set of environmental conditions. If a species is to survive there it must be adapted to these conditions.

Plants

Mangrove swamp

Mangrove swamps occur along the coasts of many countries in tropical climates. The mud in which plants grow is moved by the rising and falling of the tides.

Mangrove trees have adapted to the tides by growing many roots from their trunks. The roots spread out over a wide area and dip down into the mud to hold the tree in place. Mangrove trees also have seeds that are adapted for survival in this habitat of moving mud. When the fruit forms it remains attached to the tree. The seed germinates using moisture in the humid air, and the seedling grows to about 25 cm before it leaves the tree. As the seedling falls it remains vertical so that when it hits the mud the root forces its way in and holds the plant in place.

8 What do you think would happen to the mangrove trees if they did not have the extra roots?

9 Why is it beneficial for the seeds of the mangrove tree to germinate and the seedlings to grow while still on the tree?

Figure 3.4 A mangrove swamp. The roots are visible because it is low tide.

Tropical rainforest

The main feature of a tropical rainforest is its thick forest canopy of branches and leaves. This shades the ground below. When seeds fall to the ground and germinate, the seedlings that are produced struggle to find enough light to survive, and as a result many die. The seeds of the strangler fig are capable of growing in the compost that develops in the forks of tree branches. This means that as the seed is nearer the canopy it has a greater chance of receiving enough light to survive.

At first the seedling uses water in the compost in the fork but as it grows it needs more. The plant responds to this by growing a root down the side of the tree to find the soil on the ground. Once the root has reached the ground, it can take up more water and nutrients and the rest of the plant can start to grow upwards into the canopy. The plant continues to send down more roots, and in time they form a basket-like support around the tree. The water and minerals provided by the extra roots allow the plant to grow so large that its leaves overshadow the tree on which it is growing.

Over time, the roots develop such a rigid hold on the tree trunk that the tissues beneath the bark, which transport food and water inside the tree, are crushed and the tree dies. The strangler fig continues to thrive and takes over the space originally occupied by the tree.

10 What do you think might happen to strangler fig seeds that fall into forks of branches high in the canopy, which have very little compost?

11 How does the leafy shoot of the strangler fig affect the tree it is growing around?

Figure 3.5 A strangler fig around a tree

Keystone species

A keystone species in a habitat is one that helps the survival of a large number of other species that are adapted in some way to benefit from its presence, often through feeding.

The strangler fig is an example of a keystone species. Its fruit provides food for hornbills, monkeys, parrots, pigeons and many insects. The fruit is often produced at times when other plants are not producing fruit, so it helps to provide a constant food supply to many rainforest herbivores. If the strangler fig was removed from the forest a great many other species would suffer, and possibly become extinct.

When trees are removed from a rainforest for timber, care must be taken to consider plants like the strangler fig and preserve them. This can be particularly difficult where the strangler figs attack the trees that loggers want to cut down and sell.

For discussion

Can you work out a strategy for maintaining strangler figs in a forest while still extracting some timber by controlled logging?

Animals

Adaptations to a fast-flowing river

The tiny remains of plants and animals are washed down streams and rivers and form a food supply for any animals that are adapted to live there. The main problem for animals in such a habitat is the current, which will carry them away. Many invertebrates have solved this problem by developing ways of holding on to the river bed and presenting as small a surface as possible to the water rushing by them. Stonefly and mayfly nymphs have legs adapted for gripping rocks. Their bodies are flat and held close to the rocks so the water flows over them. Leeches have suckers to hold on to rocks. The freshwater limpet (see Figure 6.7a, page 91) has a foot that acts as a sucker and a streamlined shell to help the water flow smoothly over it.

12 Where do you think the tiny remains of plants and animals in the stream come from?

13 Why should animals become adapted to live in fast-flowing water when there are regions of slow-moving water in a river?

Adaptations to tree tops

The rainforest canopy is the habitat of many species of monkey. The monkey's body has many adaptations to tree-top life. Monkeys have small, lightweight bodies that allow them to climb out on to slender branches to collect food. Monkeys have an opposable thumb and big toe, which allow them to grip the branches firmly.

Figure 3.6 Monkey in the canopy in a tropical rainforest

14 How would the feeding activities of a monkey be affected if it had a large heavy body?

15 Although monkeys have good eyesight they also have good hearing and communicate a great deal by sound in the forest canopy. Why is this?

A monkey will often jump from one branch to another or even from tree to tree. When it does this it needs to be able to judge distance. Both eyes face forwards so that their fields of vision overlap. This allows monkeys to judge distances accurately so that they can land safely.

Monkeys have tails, which they use to help them keep their balance as they run and jump about. In South America monkeys such as the spider monkey have a prehensile tail. This acts as a fifth limb, and they use it to grip on to or even hang from branches as they feed.

Extreme adaptations

Adaptations are modifications of a basic body plan but in some species one or more extreme adaptations have developed that increase individuals' chances of survival in a habitat. These range from a great reduction in an organ to a great development of an organ or the adopting of an unusual method of movement. Here are just four examples.

Pebble plant

Pebble plants grow in the deserts of Southern Africa. They protect themselves from browsing animals by camouflaging themselves to look like stones. A pebble plant has only two leaves and these are small and close to the ground so they are difficult for browsers to reach. The two leaves make an almost spherical shape. This shape allows the plant to have a large volume for storing water, yet presents only the smallest possible surface area to the heat of the Sun's rays.

16 What might happen to a pebble plant in the heat of the Sun if it had a large, flat surface with a thin body, like a pancake?

Figure 3.7 Pebble plants are adapted for survival in harsh desert conditions.

Pit viper

The pit viper has a pit just in front of each eye, about 4 mm wide and 6 mm deep. These are packed with receptors that are sensitive to heat. The receptors are so sensitive that they can detect changes of 0.002 °C. This means that an object 0.1 °C warmer or cooler than the surroundings can be detected by the snake. These heat-sensitive organs help the pit viper to find food, such as mammals and birds, in dark places.

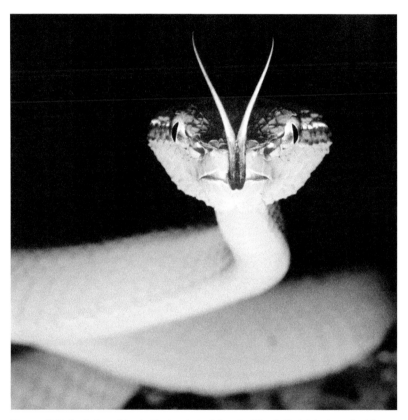

Figure 3.8 A pit viper, living in the tropical rainforest of South East Asia, 'scenting' the air with its tongue. The pit viper is adapted to hunt in the dark.

Mammals and birds are described as 'warm-blooded'. This is misleading because on a hot day, so-called 'cold-blooded' animals such as reptiles can be warmer. Mammals and birds regulate their body temperature so that it stays constant, and usually above that of the animal's surroundings. This makes birds and mammals suitable prey for pit vipers.

The areas detected by the pits on each side of the snake's head overlap just like the fields of vision of a monkey's two eyes (page 39), and helps the snake to judge the distance of its prey, so that it knows when to strike with its fangs and poison.

17 Why must the tumbleweed shoot die to spread the seeds?

18 What adaptations would a tumbleweed need so that the shoot did not have to die to spread the seeds but could plant itself again?

Tumbleweed

Plants that grow tall can disperse their seeds into the wind. The seeds may be blown a long distance before they reach the ground. Woody plants on some grasslands cannot grow tall because the winds would blow them over, so they must disperse their seeds in another way. The tumbleweed solves this problem by breaking off its shoot full of seeds – the dead shoot can then be blown over the grassland by the wind and drop its seeds as it goes.

Figure 3.9 Tumbleweed is common in desert and prairie habitats. Here, the shoot system of a mature tumbleweed has snapped off above the root and curled into a ball. As it is blown by the wind it sheds its seeds.

Flying fish

Flying fish are found in tropical seas. They feed on the plankton close to the surface, where they are the prey of the dolphin fish. Dolphin fish behave like a dolphin in some ways – for example, they can jump out of the water. When a dolphin fish starts an attack, the flying fish swims faster and faster. It moves upwards in the water, and when it is travelling at about 60 km/h it breaks through the sea's surface and glides through the air on its long, wide front fins. The lower part of the tail fin is also long, and as the fish rises into the air it waves its tail fin 50 times a second so the lower fin repeatedly pushes against the water and gives the fish extra thrust to make its flight. Once out of the water, the flying fish can travel up to 200 metres in the air and escape from the dolphin fish.

19 Do you think that the flying fish is free from predators once it has left the water? Explain your answer.

For discussion

What is your favourite animal? Discuss it with your group. If two of you have the same, select another one. When all of you have selected an animal, use books and the internet to find out how it is adapted to its habitat.

Do you have a favourite plant such as a tree or a small flowering plant such as a lily. Find out about its natural habitat and how it is adapted to it.

Figure 3.10 Flying fish have adaptations that allow them to escape their marine habitat and avoid predators.

Making close observations

When you walk through a habitat, you may see birds and insects and perhaps mammals such as squirrels. You may see the bright colours of flowers and fruits and notice the plants producing them but ignore other plants that are just in leaf. To really know about a living thing in its habitat you have to spend many days observing it at all times of year. As your observations increase, you will see how well the organism is adapted to its environment and you might make new discoveries about it. When Jane Goodall began studying chimpanzees every day in Tanzania, she made many discoveries about them.

Figure A Jane Goodall

Jane Goodall was born in 1934 in England and, from a young age, she wanted to live with animals in the wild and write about them. In 1956 she was invited to a friend's farm in Kenya and the following year, once she had saved up enough money for the trip, she arrived. After a few weeks in Africa, she met Louis Leakey (1903–1972), a Kenyan archaeologist who was greatly interested in the early forms of humans. He was impressed with Goodall's knowledge and enthusiasm for animals and believed that she could help him in his work. Chimpanzees are closely related to human species, and Leakey thought that if their behaviour could be studied closely this might provide some ideas about how very early humans behaved. He knew from the fossils of early humans that he had found that they lived on lake shores and so he hired Goodall to travel to Tanzania and study a group of chimpanzees living at Gombe on the shore of Lake Tanganyika.

Goodall arrived at Gombe by boat in 1960. Her plan was to observe the chimpanzees daily using binoculars, make notes and write up a report. She aimed to have the project finished in four months. At first, as she entered the forest at the lake shore, the chimpanzees ran away when they saw her at a distance of →

about 500 metres. Goodall was patient and she tried again and again to approach them. Eventually, after some months, her perseverance was rewarded and the chimpanzees stopped running away and even allowed her to approach them. She was allowed to get so close that she could clearly see their faces. At the time, scientists studying individual animals in their habitats gave them numbers but Goodall thought differently. She gave them names.

From the evidence of other observers, chimpanzees were known to have opposable big toes just as we have opposable thumbs to help us grip. With chimpanzees, the big toes allow them to grip branches as they swing from tree to tree. A chimpanzee's long arms and strong muscles also help it move through the forest. As Goodall continued with her observations she made more discoveries about how chimpanzees were adapted to their environment.

Chimpanzees were thought to live in chaos, in groups with no real organisation, and to be mainly herbivorous – their large, fang-like canine teeth were believed to be just for display, to show aggression. Goodall found that one day the chimpanzees were eating the meat from a bush pig. Some time later she observed them hunting, catching and killing a red colobus monkey. When chimpanzees catch and kill animals, others in the group make begging gestures for food and the meat is shared out. Chimpanzees also communicate with each other by expressions on their faces and by making different calls. This helps the group to live in an orderly way.

At that time, it was believed that humans were the only animals to use tools – a tool is a device that is used to make a task easier. One day, Goodall saw a chimpanzee called David Greybeard poking grass stalks into a termite mound and then raising them to his mouth. After he had gone away, Goodall took a grass stalk and poked it into the termite mound just as the chimpanzee had done. When she pulled it out, she discovered termites biting the stalk so hard that they stuck to it. Jane believed the chimpanzee was eating the termites when they were raised to his mouth. The grass stalk was being used as a tool to obtain food. Shortly after this observation, Goodall saw David Greybeard and other chimpanzees take twigs, strip their leaves off and use them like the grass stalks to feed on termites. The chimpanzees were not only using tools but also making them.

Jane Goodall's work continues today at the research centre she set up at Gombe in 1965.

1 What evidence from his work did Louis Leakey use in selecting a place for Jane Goodall to study chimpanzees?

2 What equipment did Goodall use to make and record her observations?

3 What signs of a scientist did Goodall display as she tried to observe the chimpanzees?

4 What example of creative thought did Goodall display in recording the actions of the chimpanzees?

5 How did Goodall's observations change our knowledge of chimpanzees?

For discussion

Find out about the work of other scientists living today, using books and the internet. In your group, you could select different universities and find out about the work of the scientists there. After finding the university website, look for the staff directory where the research interests of the scientists may be displayed. Report back to the group with your findings.

◆ SUMMARY ◆

◆ Living things in a habitat are adapted to the seasons (*see page 35*).
◆ Plants are adapted to the environmental conditions in their habitat (*see page 37*).
◆ Animals are adapted to the environmental conditions in their habitat (*see page 39*).
◆ Some plants and animals have greatly adapted body parts for survival in their habitat (*see page 40*).
◆ Jane Goodall is an example of a scientist who studies the natural world (*see page 43*).

End of chapter question

Below is a description of a fictional habitat for which you must design an animal with adaptations to take advantage of the changing food supply.

In the habitat is a large lake, rich in small invertebrates. Around the edge are tall trees, which produce leaves only in the top branches. Once a year the lake dries up and at this time caterpillars hatch and feed on the leaves. After a few weeks they change into flying insects. The flying insects spend most of the following weeks either in the air or resting on the branches. They die when the rains return and the lake fills up again.

You might like to start by thinking about an animal swimming in the lake. It could be large or small, and could have changes in its life cycle, like a frog.

4 Ecosystems

- The growth of ecology
- A vocabulary of ecology
- Food chains
- Food webs
- Ecological pyramids
- Decomposers
- Ecosystems
- How populations change

The growth of ecology

When you take a walk in a park or the countryside, some plant or animal may make you stop and look at it. It could be the colour or shape of a flower that attracts you or the activity of an animal such as a bird.

About 250 years ago, explorers on voyages of discovery from European countries such as England, Portugal and Spain, did the same thing as they cut their way through rainforests, trekked across plains and climbed mountains. Some of the explorers examined the living things more closely and considered how they were adapted to their surroundings just as you did when studying Chapter 3.

Alexander von Humboldt (1769–1859), a German explorer, made many journeys and from the data he

Figure 4.1 Alexander von Humboldt explored many new habitats in South America.

collected he noticed that the number of species living in a place increased as he moved from the north to the tropics. The idea of different worldwide habitats began to form.

Philip Sclater (1829–1913) was an English biologist who looked at the distribution of song birds, such as the sparrow, around the world and decided that the Earth could be divided up into regions based on the birds found there. These regions are known as biogeographical realms.

Alfred Russell Wallace (1823–1913), a Welsh biologist who had travelled widely, found that other animals he had seen also fitted into Sclater's system of realms. The system continued to be developed and a modified version of it is still in use today (Figure 4.2).

While the idea of realms continued to develop, a German zoologist called Karl Möbius (1825–1908) investigated oyster banks on the German coastline to see how they may be farmed. From the results of his investigations he showed how living things interact with each other in their community.

1 Which biogeographical realm do you live in?

2 Which biome in that realm do you live in?

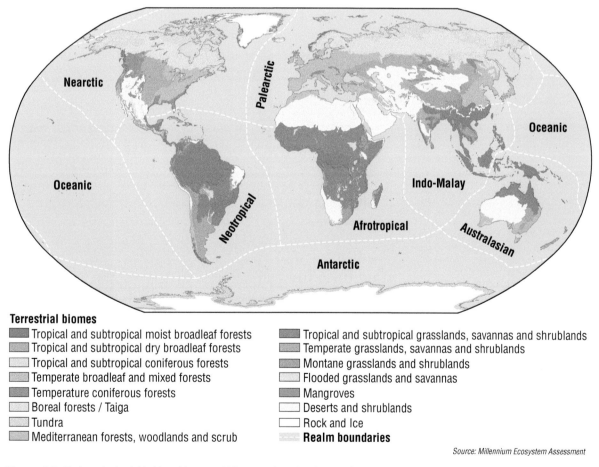

Terrestrial biomes

- Tropical and subtropical moist broadleaf forests
- Tropical and subtropical dry broadleaf forests
- Tropical and subtropical coniferous forests
- Temperate broadleaf and mixed forests
- Temperature coniferous forests
- Boreal forests / Taiga
- Tundra
- Mediterranean forests, woodlands and scrub
- Tropical and subtropical grasslands, savannas and shrublands
- Temperate grasslands, savannas and shrublands
- Montane grasslands and shrublands
- Flooded grasslands and savannas
- Mangroves
- Deserts and shrublands
- Rock and Ice
- **Realm boundaries**

Source: Millennium Ecosystem Assessment

Figure 4.2 Each realm is divided into biomes, which are regions that have a distinct community of plants and animals and a certain type of climate.

3 What pattern did von Humboldt see in his data?

4 What pattern did Sclater see in the data he studied?

5 What evidence from a secondary source and evidence from first-hand experience did Wallace use to help the development of the map shown in Figure 4.2?

6 What was the idea that started Möbius's investigations?

The idea of interaction gathered pace and Ernst Haeckel (1834–1919), a German biologist, devised the word 'ecology' in the 1860s to describe the relationships between animals, other living things in the environment and the features of the environment such as the weather and the type of soil. In the following decades, interest in such relationships gradually grew and by the beginning of the 20th century, **ecology** – the study of living things in their environment – was established as a science.

A vocabulary of ecology

There are a number of special terms that ecologists use frequently so it is important to know what they mean. Here is a list of the most frequently used words and phrases, to give you a quick reference to ecology vocabulary as you study this chapter.

- **Habitat** – the place where a plant, animal or microorganism lives.
- **Community** – all the plants, animals and microorganisms that live in a habitat.
- **Environment** – the surroundings of a living thing. A complete description of an environment includes all the living things, rocks, soil and weather conditions.
- **Abiotic factors** – factors due to the non-living part of an environment, such as weather, rocks and soils, which may affect the survival of a living thing.
- **Biotic factors** – factors in an environment due to living things, such as amount of food or number of predators, which may affect the survival of a living thing.
- **Ecosystem** – the community of living things and the abiotic factors found in a particular place, such as a woodland or a pond.
- **Biodiversity** – a term used to describe the number and variety of species in an ecosystem.
- **Biome** – a large region of the Earth generally covered by the same community of plants and animals and all parts having the same weather.
- **Herbivore** – an animal that only eats plants.
- **Carnivore** – an animal that only eats other animals.
- **Omnivore** – an animal that eats both plants and animals.
- **Predator** – an animal that feeds or preys on another animal.
- **Prey** – an animal that is eaten by or falls prey to a predator.

- **Producer** – an organism that produces food at the beginning of a food chain (usually a plant).
- **Consumer** – an animal that eats plants, other animals or both.
- **Primary consumer** – an animal that eats plants (this may be a herbivore or an omnivore).
- **Secondary consumer** – an animal that eats primary consumers (this may be a carnivore or an omnivore).
- **Tertiary consumer** – an animal that eats secondary consumers (this may be a carnivore or an omnivore).
- **Quaternary consumer** – an animal that eats tertiary consumers.
- **Top carnivore** – the animal at the end of the food chain.
- **Food chain** – a description (often a diagram) of the way some organisms in a habitat are linked to each other through feeding. A food chain begins with a producer and is followed by one or more consumers, for example:

    grass → gazelle → lion

- **Food web** – a description (often a diagram) of how a number of food chains in a habitat are linked together to show how food and energy pass through the habitat.

Food chains

Food chains in the ocean

The animal shown in Figure 4.3 is a killer whale. It is an organism at the end of a long food chain. The killer whale is trying to catch a seal. The seals, sensing that the killer

Figure 4.3 A killer whale hunting a seal

whale is close by, have come out of the water. This is not stopping the killer whale from pursuing them – it has swum quickly up the shore and tried to catch a seal by surprise in the surf.

In time, the seals will have to return to the sea to catch food for themselves. They feed on a variety of large fish that live off the shore. The large fish in turn feed on smaller fish – sometimes fish of their own species. The smaller fish feed on the larvae of crustaceans such as crabs and lobsters. These larvae do not remain on the rocky bed of the sea like their parents, but swim in the sunlit open waters of the sea. They form part of the plankton – a massive collection of tiny animals and algae. The crab larvae, along with all the other tiny animals, eat the algae. The algae do not eat anything. They are plant-like microorganisms belonging to the Protoctista kingdom. Like plants, algae make their own food through photosynthesis, and they use energy in sunlight to do it. It is the energy in sunlight that was trapped in algae and passed up the food chain that is being used by the killer whale in Figure 4.3 as it tries to catch another meal.

Food chains around the world

There are food chains in every habitat in the world. Most are not as long as the one you have made in answer to question **7**. Food chains have a particular structure. They begin with an energy source, which is usually the Sun. The first organism in the food chain is the one that extracts energy from sunlight and uses it to produce food.

These organisms are called producers. Any green plant such as a moss, fern, conifer or flowering plant is a producer. Algae are producers too.

An organism that feeds on a producer is called a primary consumer. It in turn is eaten by a secondary consumer, which in turn is eaten by a tertiary consumer, and so on.

Because of their similar structure, we can compare food chains from around the world.

For example, in the Australian outback, which is an area of desert and grassland, the main producer is grass. A primary consumer is a kangaroo and a secondary consumer is a wild dog (dingo).

7 a) Read the text in this section again and construct a food chain from the information.

b) Describe the path of energy through the food chain. Remember that energy is used by living things for life processes such as movement and use this knowledge in your answer.

8 Describe each organism in the food chain using ecological terms such as 'producer' and 'consumer'.

Figure 4.4 A female kangaroo with a joey in her pouch feeding on grass. She is a primary consumer.

In a tropical rainforest, there are many different kinds of plants that are producers. For example, a tree is a producer and the beetle that feeds on its leaves is a primary consumer. A tree frog feeds on the beetle and is therefore a secondary consumer. A tree snake feeds on the frog and is a tertiary consumer.

Figure 4.5 This frog is eating a damselfly. Here, the frog is a secondary consumer.

In the mountains of Europe, heather covers large areas of the ground. It is a producer. Mountain hares nibble its shoots and so are primary consumers. The golden eagle feeds on mountain hares and is a secondary consumer.

Figure 4.6 A golden eagle with its prey. The eagle is a secondary consumer and, because it has no predators, it is called a top carnivore.

The energy in food chains

When you eat a radish or a carrot, for example, you eat most of the plant except the leaves. Even if you were to eat a whole plant, you would not take in all the energy that the plant had trapped to make food. While the plant was growing it used some of the energy to stay alive and to build new materials so that its cells could grow and divide.

In a similar way, if you were to be eaten by a predator it would not receive all the energy that you had taken in from your food. You use energy for life processes such as breathing, digesting your food and moving your body. You even lose some energy as heat to your surroundings. (In fact all living things lose some heat to their surroundings owing to the chemical reactions that take place inside them to keep them alive.)

This means that as energy passes along a food chain some energy is lost from the food chain at every link.

Where are you in your food chains?

Look at Figure 4.7. Where do all these foods come from? The fruit comes from plants and the juice comes from plants too. Marmalade is made from oranges. Eggs are laid by hens which feed on cereal grains such as maize. A slice of bread is made from the flour of another cereal called wheat, and it is spread with margarine made from sunflower oil. Cheese is made from milk, which comes from cows, and the cow gets the materials and energy it needs to make milk from the grass it eats. Breakfast cereals are made from maize, oats and wheat.

12 Write a food chain for each item in the meal in Figure 4.7.

13 Write down the items in your breakfast. Make a food chain from each of these items. How many food chains feature animals as well as plants?

14 When do you feed as:
   a) a primary consumer
   b) a secondary consumer?

Figure 4.7 A selection of breakfast foods

Food webs

When several food chains are studied in a habitat, some species may appear in more than one. For example, a panda eats bamboo shoots and also eats fish. The two food chains it appears in are:

$$bamboo \rightarrow panda$$

$$water\ plants \rightarrow snails \rightarrow fish \rightarrow panda$$

These food chains can be linked to make a food web. A food web shows the movement of food through a habitat. It can also be used to help predict what might happen if one of the links in a food web was absent.

Look at Figure 4.9 and think of each animal shown there not just as one animal but as the whole **population** of that species of animal in the wood. If you think of each animal in the plural, such as 'voles' and 'finches', it may help you think about animal populations.

Now imagine that the trees had a disease that made their leaves fall off. The caterpillars would starve and die, and they would not be available as food for the robins. This means that the robins must eat more beetles and woodlice if they are not to go hungry. The reduction in the number of woodlice would affect the shrews, as this is their only food. The beetle population would also fall, forcing the foxes to search out more voles to eat.

Figure 4.8 Pandas eat bamboo, but they also eat fish, so they appear in more than one food chain. Animals that eat plants and animals are called omnivores.

15 How might the populations of other species in the woodland change if each of the following was removed in turn?
a) foxes
b) seeds

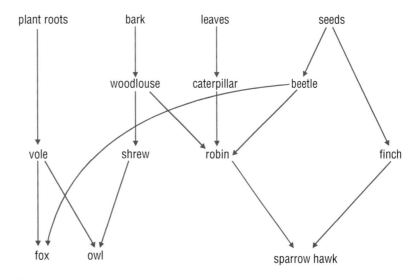

Figure 4.9 Example of a food web in a European woodland

Ecological pyramids

When a survey of a habitat is complete, ecologists – scientists who study ecosystems – can examine the relationships between the different species. They may find, for example, that deer depend on ferns for shelter in a wood, and that some birds use moss to line their nests. The major relationship between organisms in a habitat is the relationship through feeding. It is this relationship that interests ecologists most.

By studying the diets of animals in a habitat, ecologists can work out food chains and ecological pyramids.

Pyramid of numbers

16 What would happen to the number of rabbits and grass plants if the number of foxes:
a) increased
b) decreased?

17 What would happen to the number of grass plants and foxes if the number of rabbits:
a) increased
b) decreased?

The simplest type of ecological pyramid is the pyramid of numbers. The number of each species in the food chain in a habitat is estimated. The number of plants may be estimated using a quadrat. The number of small animals may be estimated by using traps, nets and beating branches. The number of larger animals, such as birds, may be found by observing and counting.

An ecological pyramid is divided into layers or tiers. There is one tier for each species in the food chain. The bottom tier is used to display information about the plant species or producer. The second tier is used for the primary consumer and the tiers above are used for other consumers in the food chain. The size of the tier represents the number of the species in the habitat. If the food chain below is represented as a pyramid of numbers, it takes the form shown in Figure 4.10.

grass → rabbit → fox

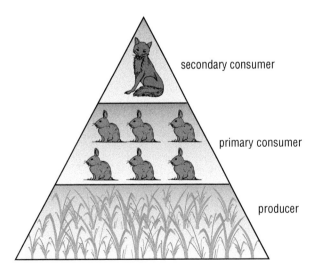

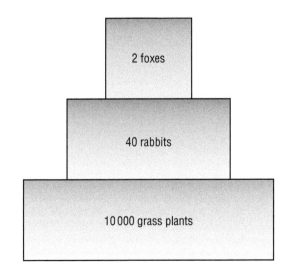

Figure 4.10 Pyramid of numbers of grass plants, rabbits and foxes

18 Why do the two food chains considered here produce different pyramids of numbers?

19 Why do you think there are usually more organisms at the bottom of a food chain?

Not all pyramids of numbers are widest at the base. For example, a tree creeper is a small brown bird with a narrow beak that feeds on insects, which in turn feed on an oak tree. The food chain of this feeding relationship is:

oak tree → insects → tree creeper

When the food chain is studied further and a pyramid of numbers is displayed, it appears as shown in Figure 4.11.

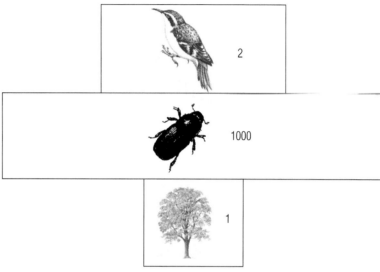

20 Living things need water in their bodies to survive. What happens to the living things used to work out a pyramid of biomass? Explain your answer.

21 oak tree → insects → tree creeper

If you drew a pyramid of biomass for this food chain, what do you think it would look like? How would it compare with the pyramid of numbers? Explain any differences that you would see.

Figure 4.11 Pyramid of numbers of oak tree, insects and tree creepers

Pyramid of biomass

The amount of matter in a body is found by drying it to remove all water, and then using scales to find the mass. This amount of matter is called the **biomass**. Ecologists find measuring biomass useful as it tells them how much matter is locked up in each species of a food chain.

Decomposers

Not only do the living bodies of each species provide food for others but their dead bodies and waste are food too. The dead bodies of plants and animals are food for fungi, bacteria and small invertebrates that live in the soil and leaf litter. These organisms are called **decomposers**. When they have finished feeding, the bodies of plants and animals become reduced to the substances from which they were made. For example, the carbohydrates in a plant are broken down to carbon dioxide and water as the decomposers respire. Other substances are released from the plant's

22 Why are decomposers important? How do they affect you?

body as minerals and return to the soil. Decomposers are recyclers. They recycle the substances from which living things are made so that they can be used again.

Figure 4.12 These Malaysian termites are feeding on leaf litter.

Ecosystems

Decomposers form one of the links between the living things in a community and the non-living environment. Green plants form the second link. When a community of living things, such as those that make up a woodland, interact with the non-living environment – the decomposers releasing minerals, carbon dioxide and water into the environment and then plants taking them in again – the living and non-living parts form an ecological system or ecosystem. An ecosystem can be quite small, such as a pond, or as large as a lake or a forest.

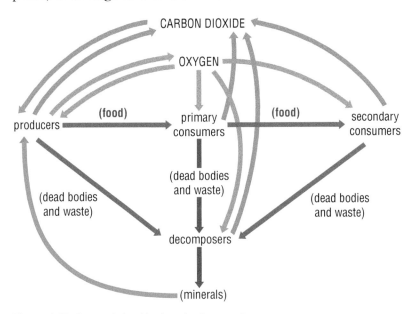

Figure 4.13 Some relationships in a simple ecosystem

23 How might studying ecosystems help to conserve endangered species?

24 Look at the ecosystem set up in the aquarium tank in Figure 4.14 and answer these questions.
  a) Which are the producers and which are the consumers?
  b) i) Construct some food chains that might occur in the tank.
    ii) For each food chain, explain how energy passes along it.
  c) Where are the decomposers?
  d) Give examples of some ways in which the living things react to their non-living environment.

Working out how everything interacts is very complicated but is essential to ecologists if they are to understand how each species in the ecosystem survives and how it affects other species and the non-living part of the ecosystem. Figure 4.13 shows how the living and non-living parts of a very simple ecosystem react together.

The ecosystem in an aquarium tank

A pond life aquarium is a very easy ecosystem to study as it can be set up inside and observed at any time and in any weather. The glass walls allow you to get close to the living things to observe them.

Figure 4.14 shows an aquarium containing a pond ecosystem on a small scale. The fish are sticklebacks, which are found in North America, China, Europe and Japan.

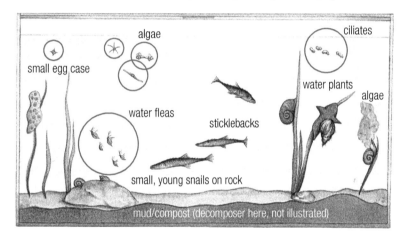

Figure 4.14 An aquarium with pond life. The circled organisms are greatly magnified.

How populations change

If an area of ground is cleared of vegetation, it will soon be colonised by new plants and animals. The following is an account of how an area of soil could be colonised. The colonisation described is much simpler than what would occur naturally so that the ways in which the plants and animals interact can be seen more clearly.

A seed lands in the centre of a soil patch and germinates. The plant is an ephemeral (which means it has a very short life cycle) so it is soon fully grown and producing flowers and seeds. The seeds are scattered over the whole area. They all germinate and grow so the population of ephemeral plants increases.

Outside the area of cleared soil there are perennial plants (which flower and continue to grow for several years). These have stems with broad leaves that cover the

25 In what ways do the two kinds of plants compete for the resources?

26 If the herbivorous insects had not arrived what do you think would have happened to the two populations of plants?

27 How did the arrival of the herbivorous insects affect your prediction in question **26**? Explain your answer.

28 What effect do the carnivorous insects have on the population of:

a) herbivorous insects

b) ephemeral plants?

ground. As the population of ephemeral plants increases, the perennial plants grow into the area of cleared soil. The two kinds of plants compete for light, water and minerals in the soil. As the numbers of both plants increase, the competition between them also increases. The perennials compete more successfully than the ephemeral plants for the resources in the habitat and produce more offspring. The broad leaves of the perennial plants cover the soil and prevent seeds from landing there and germinating. The leaves may also grow over the young ephemeral seedlings. In time the ephemeral plants that are producing seeds will die and the stems and broad leaves of the perennials could cover them too.

Herbivorous insects land on some of the perennial plants and start to feed on their leaves. They feed and breed and as their numbers increase they spread out over other perennial plants in the area. The population of the perennial plants in the area begins to fall and the population of ephemerals, which are not eaten by the insects, begins to rise.

A few carnivorous insects land in the centre of the patch. They clamber about on the plants and feed on the herbivorous insects. The well-fed carnivorous insects breed and their population increases. The population of the herbivorous insects starts to fall.

How populations change with time

A population of a species is the number of individuals present in a habitat. The survival of a species in a habitat depends upon many factors. As prey animals are affected by predators and predators depend on prey, it seems reasonable to suspect that their population sizes might be linked.

Charles Elton (1900–1991) was an English ecologist who studied some unusual data to investigate the relationship between prey and predator. He used the records of the Hudson Bay Company in Canada for the years 1845 to 1935. The Hudson Bay Company traded in animal pelts and the records showed the numbers of pelts supplied to the company by trappers each year. Elton studied the numbers of snowshoe hare pelts supplied by the trappers as well as the numbers of pelts of the lynx, which is a predator of the snowshoe hare. The number of pelts of each species supplied in each year indicated the size of the species population in that year. Figure 4.15 shows the graphs produced from the data.

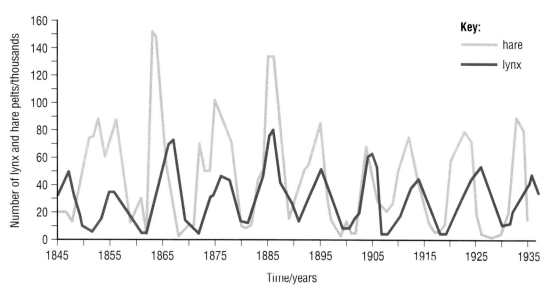

Figure 4.15 The changes in the numbers of animal pelts supplied by trappers to the Hudson Bay Company

29 a) How does the population of snowshoe hares vary over the 90 years?

b) How does the population of lynx vary over the 90 years?

c) Do the populations vary in exactly the same way or does one population lag behind the other?

d) It has been suggested that the way the population varies is due to the way the lynx preys on the snowshoe hare. Do you think this is an accurate suggestion? Explain your answer.

30 a) Think about the colonisation described on pages 57 and 58. How do you think the population of herbivorous insects would change over a longer period of time?

b) Sketch a freehand graph to show how the population of herbivorous insects would change over time.

c) On the same axes, sketch a graph to show how the population of carnivorous insects would change over time.

Predicting changes in populations

The changing size of the human population can be predicted by comparing birth rates with death rates. The birth rate is the number of babies born per 1000 people in the population in a year. The death rate is the number of people dying per 1000 people in the population in a year.

If the birth rate is greater than the death rate, the population will increase in size. If the death rate is greater than the birth rate, the population will decrease in size. If the birth rate and the death rate are the same the population will remain unchanged.

Table 4.1

Date	Population (billions)
1804	1
1927	2
1960	3
1974	4
1987	5
1999	6
2013	7
2028	8
2054	9
2183	10

For discussion

Large mammals, like the Javan rhinoceros, need large areas of natural habitat to support a large population. With the increasing human population, why is it difficult to conserve these large areas? Explain your answer.

31 If the estimates are correct, how old will you be when the world population reaches:
   a) 7 billion
   b) 8 billion
   c) 9 billion?

32 Plot a graph of the data in Table 4.1. What might the population be in 2100?

Birth rates, death rates and conservation

Many endangered mammal species have been reduced to a small world population by hunting. The animals have been killed more quickly than they can reproduce. If the death rate exceeds the birth rate, the species is set on a course for extinction. Many mammals are now threatened with this course.

They can be helped by raising their birth rate and reducing their death rate. Zoos can help increase the size of the world population of some endangered animal species. They increase the birth rate by ensuring that all the adult animals in their care are healthy enough to breed and by providing extra care in the rearing of the young. Zoos also reduce the death rate by protecting the animals from predation. In many countries reserves have been set up in which endangered animals live naturally but are protected from hunting by humans. This reduces the death rate, which in turn increases the birth rate as more animals survive to reach maturity and breed.

Figure 4.16 The Javan rhinoceros was once numerous throughout Southeast Asia, but now only one small population of under 50 animals is known for certain, in a protected area in Indonesia. The species is critically endangered and on the brink of extinction.

The human population

The human population of the Earth reached 1 billion (a thousand million) in 1804. Two hundred years later the figure was over six times greater. Table 4.1 shows the rise in population to 6 billion, and the estimated population in the future.

For discussion

Why do you think the human population has grown so large?

Habitat surveys

People have been surveying habitats since the earliest times. The purpose was to find fruit, seeds and roots, as well as animals such as mammals and birds, to eat. Much later, when countries in Europe sent out sailing ships to explore the world, habitat surveys were made at each new land that was visited. They were not called habitat surveys. They were just journeys through the new land where the first explorers, often with no scientific knowledge, observed and collected specimens. The explorers wrote down their observations and brought specimens home, by which time they were usually dead and in some form of decay. However, the arrival of specimens from new lands fired the interest of scientists and eventually the expeditions became more scientific. Artists took part to draw specimens in their habitats, and scientists recorded observations more systematically in notebooks and preserved dead specimens in bottles of alcohol for the journey home.

In the 1950s, television companies sent out expeditions in which the plant and animals of a habitat were recorded on film. Wildlife films like this are still being made today and stimulate many people to take up a career in biology.

There is a second type of habitat survey taking place today which is much more scientific. The aim of these surveys is to describe the habitat as accurately as possible for two reasons. First, the habitat might be that of a rare or endangered species so the purpose is to discover how the activities of the species link with other species and so work out the best way of setting up a conservation programme. Second, the habitat may be one that is under threat from the human population and the land on which it stands is being considered for building houses and factories or for flooding to make a reservoir for a hydroelectric power station.

So how are habitat surveys made today? Here are the stages that you would go through if you were a scientist carrying out such a survey.

The chances are that even the most remote habitat has been visited by scientists in the past and there may be records of their observations that can be examined. This process of looking at recorded data and accounts of previous journeys in the habitat is called the ecological desk survey.

1 When the earliest people surveyed a habitat for plants, how do you think they might have categorised the species they found based on the evidence they observed?

2 When the earliest people surveyed a habitat for animals and saw them moving around, how do you think they might have categorised the different species based on the evidence they observed?

3 How reliable was the evidence from the first expeditions? Explain your answer.

4 How reliable was the evidence from the later expeditions? Explain your answer.

For discussion

What television programmes about wildlife have you seen recently? How good are they at providing information about habitats and the lives of plants and animals? Explain your answer. In what ways might the programmes make people concerned about they see?

Figure A These scientists are using an insect trap as part of a habitat survey in Indonesia.

Next the scientists visit the habitat and make a survey in which they note down its features and make a map. They may also talk to people who live locally about the plants and animals they have seen. Often these local people have a different language from the scientists and may have several names for a plant or an animal.

After the survey, transects are set up over the habitat. Along each transect, at regular intervals, stations are set up at which the scientists record what they can see and hear. They assess the numbers of each species at the station using the DAFOR code – the letters stand for Dominant, Abundant, Frequent, Occasional, Rare. On the walk out along the transect the abundance of the different species of animals might be recorded along with their activities. On the return walk the abundance of plants might be noted together with their condition – for example, the plants might be 'in flower' or 'bearing fruits'. If there is a team of scientists each one may walk a transect at different times in the day.

Night surveys are also taken – this is when the scientists walk the transects in darkness and note the animals heard or seen by torchlight. A bat recorder may be switched on to record the sounds of bats, which humans cannot hear. If insect traps have been set up at the stations, these can be examined. At some stations, photo traps can be set up in which a camera triggered by a harmless trip wire records an image of any animal that passes by after the scientists have gone.

Surveys along these lines are taking place all over the world right now. One example is along the Kenyan coast were rainforest surveys are being made in the study of black and white colobus monkeys, which are an endangered species.

Figure B Information about the habitat of black and white colobus monkeys in the Kakamega Forest Reserve in Western Kenya could help in the protection of this vulnerable species.

5 Could the desk survey provide inaccurate evidence? Explain your answer.

6 How accurate do you think evidence given by local people might be? Explain your answer.

7 How do you think the evidence provided by using a transect for one day and one night might be made more reliable?

8 Make a plan for a local habitat survey, perhaps part of your school grounds, to describe the habitat as accurately as possible.

◆ SUMMARY ◆

◆ Scientists have worked in the past to build up the science of ecology (*see page 46*).

◆ There are many specific terms used in ecology (*see page 48*).

◆ Food chains are found in all kinds of habitats around the world (*see page 49*).

◆ Energy is passed through a food chain (*see page 51*).

◆ Humans are part of many food chains (*see page 52*).

◆ Food chains link together to form food webs (*see page 52*).

◆ A pyramid of numbers shows the numbers of organisms of each species in each layer in a food chain (*see page 54*).

◆ Decomposers break down dead bodies and wastes of living things into simple substances (*see page 55*).

◆ The living and non-living parts of a habitat form an ecosystem (*see page 56*).

◆ Populations of species in a habitat change over time (*see page 58*).

◆ Birth rates and death rates are important in predicting changes in populations (*see page 60*).

◆ A survey to estimate populations in a habitat can be made using the methods of present day scientists (*see page 61*).

End of chapter questions

1 Construct a food web for the African plains from the following information.

Giraffes feed on trees.

Elephants feed on trees.

Eland feed on trees and bushes.

Hunting dogs feed on eland and zebra.

Finches feed on bushes.

Mice feed on roots.

Baboons feed on roots and locusts.

Gazelles, zebras and locusts feed on grass.

Foxes feed on mice.

Lions feed on eland, zebra and gazelle.

Hawks feed on finches

Eagles feed on baboons, foxes and gazelles.

You may like to write the name of each living thing on a separate card and arrange the cards on a sheet of paper. You could write arrows on the paper between the cards but use pencil at first as you may find that you have to move the cards about to make the food web tidy.

2 Use the food web you have made for question **1** to answer these questions.

a) Which living organisms are producers?

b) Which living organisms are herbivores?

c) Which living organisms are carnivores?

d) An omnivore is an animal that feeds on plants and animals. Which living organism is an omnivore?

3 An investigation on the populations of two microorganisms was carried out.

One microorganism, called paramecium (a ciliate), was the predator and the other, yeast, was the prey. The microorganisms were kept in a sugar solution in a glass container. The sugar provided food for the yeast. Figure 4.17 shows how the populations varied over a period of 19 days.

a) When the population of yeast rose from its smallest to its largest, what happened to the population of paramecium?

b) Suggest a reason for your answer to part **a)**.

c) When the population of yeast fell from its largest to its smallest, what happened to the population of paramecium?

d) Suggest a reason for your answer to part **c)**.

e) Do you think the results of this experiment provide more reliable evidence for the relationship between prey and predator than the data provided by the Hudson Bay Company (pages 58–59)? Explain your answer.

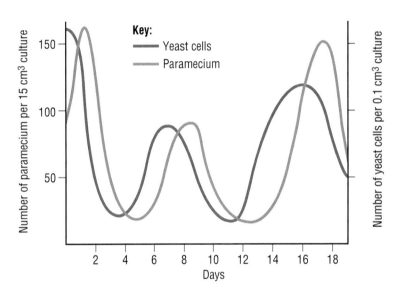

Figure 4.17

5 Human influences on the environment

◆ The growing impact of humans on the environment
◆ The Earth's changing atmosphere
◆ Air pollution
◆ The Gaia Hypothesis
◆ Water pollution
◆ Indicators of pollution
◆ Intensive farming
◆ Biological pest control

Humans in the environment

For discussion

What metal products do we use in our daily lives? How long can you make the list?

The first people used natural materials such as stone, wood, animal skins, bones, antlers and shells. They shaped materials using flint knives and axes. When they discovered fire they also discovered the changes that heat could make. First they saw how it changed food, and later it is believed that they saw how metal was produced from hot rocks around a camp fire. In time, they learned how to extract metals from rocks by smelting and to use the metals to make a wide range of products (Figure 5.1).

The human population was only small when metal smelting was discovered, and the smoke and smell from this process caused little pollution. As the human population grew, the demand for metal and other products such as pottery and glass increased. All the processing in the manufacture of these products had to be done by hand. Although there would be some pollution around the places where people gathered to make these products, the world environment was not threatened.

About 200 years ago it was discovered how machines could be used in manufacturing processes, and the Industrial Revolution began. Machines could be used to produce more products than would be produced by people working on their own.

This meant that large amounts of fuel were needed to work the machines and air pollution increased (Figure 5.2). Larger amounts of raw materials were needed and more habitats were destroyed in order to obtain them. More waste products were produced, increasing water and land pollution as the industrial manufacturing processes developed. The world population also increased, causing an increased demand for more materials, which in turn led to more pollution and habitat destruction.

Figure 5.1 A Bronze Age village scene in Northern Europe, around 4000 years ago. In the foreground, a furnace is being used to smelt metal, while elsewhere metal tools and objects are in use.

Figure 5.2 The smoky skyline of Glasgow in the mid-19th century

1. What were the first materials people used?
2. Why did the pollution caused by manufacturing materials not cause a serious threat to the environment until the Industrial Revolution?
3. Why did people believe it was safe to release wastes into the environment?

At first, and for many years, it was believed that the air could carry away the fumes and make them harmless, and that chemicals could be flushed into rivers and the sea where they would be diluted and become harmless. Also, the ways various chemical wastes could affect people were unknown.

An awareness of the dangers of pollution increased in the latter half of the 20th century, and in many countries today steps are being taken to control it and develop more efficient ways of manufacturing materials.

The Earth's changing atmosphere

Studies from astronomy and geology have shown that the Solar System formed from a huge cloud of gas and dust in space. The Earth is one of the planets formed from this cloud. The surface of the Earth was punctured with erupting volcanoes for a billion years after it formed. The gases escaping from inside the Earth through the volcanoes formed an atmosphere composed of water vapour, carbon dioxide and nitrogen.

Figure 5.3 Gases escaping from volcanoes, like this one in Indonesia, formed the Earth's first atmosphere.

Figure 5.4 Some early land plants were probably similar to modern-day ferns.

Three billion years ago the first plants developed. They produced oxygen as a waste product of photosynthesis. As the plants began to flourish in the seas, in fresh water and on the land, the amount of oxygen in the atmosphere increased. It reacted with ammonia to produce nitrogen.

Bacteria developed that survived by using energy from the breakdown of nitrates in the soil. In this process, more

4 How has the composition of the atmosphere changed since the Earth first formed?

5 What has changed the composition of the atmosphere?

For discussion

How is the change in the ozone layer affecting people today?

For discussion

How would our lives change if power stations could no longer supply us with electricity?

nitrogen was produced. In time, nitrogen and oxygen became the two major gases of the atmosphere.

Between 15 and 30 kilometres above the Earth, the ultraviolet rays of the Sun reacted with oxygen to produce ozone. An ozone molecule is formed from three oxygen atoms. It prevents ultraviolet radiation, which is harmful to life, reaching the Earth's surface. If the ozone layer had not developed, life might not have evolved to cover such large areas of the planet's surface as it does today.

Owing to the activities of humans, the atmosphere today contains increasing amounts of carbon dioxide, large amounts of sulfur dioxide, and chlorofluorocarbons (CFCs), which have destroyed large portions of the ozone layer.

Air pollution

Every day we burn large amounts of fuel such as coal and oil in power stations to produce electricity. This provides us with light, warmth and power. The power is used in all kinds of industries for the manufacture of a wide range of things, from clothes to cars. In the home, electricity runs washing machines, refrigerators and microwave ovens. It provides power for televisions, radios and computers. When coal and oil are burned, however, they produce carbon dioxide, carbon monoxide, sulfur dioxide, oxides of nitrogen and soot particles that make smoke. When polluting substances are given out into the air in this way they are often called emissions.

Figure 5.5 Electricity makes our lives more comfortable.

Carbon dioxide

Carbon dioxide is described as a 'greenhouse gas' because the carbon dioxide in the atmosphere acts like the glass in a greenhouse. It allows heat energy from the Sun to pass through it to the Earth, but prevents much of the heat energy radiating from the Earth's surface from passing out into space. The heat energy remains in the atmosphere and warms it up (Figure 5.6). The warmth of the Earth has allowed millions of different life forms to develop, and it keeps the planet habitable.

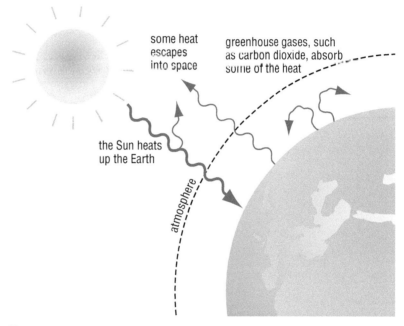

some heat escapes into space

greenhouse gases, such as carbon dioxide, absorb some of the heat

the Sun heats up the Earth

atmosphere

Figure 5.6 Some of the heat that the Earth receives from the Sun is trapped in the atmosphere.

In the past the level of carbon dioxide in the atmosphere has remained low, but the level is now beginning to rise. The extra carbon dioxide will probably trap more heat energy in the atmosphere. A rise in the temperature of the atmosphere will cause an expansion of the water in the oceans. It will also cause the melting of the ice caps at the North and South Poles and this water will then flow into the expanding ocean waters. Both of these events will lead to a raising of the sea level and a change in the climate for almost all parts of the Earth. The rise in temperature is known as **global warming**.

Acid rain

Sulfur dioxide is produced by the combustion of sulfur in a fuel when the fuel is burned. Sulfur dioxide reacts with water vapour and oxygen in the air to form sulfuric acid. This may fall to the ground as acid rain or snow.

Oxides of nitrogen are converted to nitric acid in the atmosphere and this also falls to the ground as acid rain or snow.

When acid rain reaches the ground it drains into the soil, dissolves some of the minerals there and carries them away. This process is called **leaching**. Some of the minerals are needed for the healthy growth of plants. Without the minerals the plants become stunted and may die.

The acid rain drains into rivers and lakes and lowers the pH of the water. Many forms of water life are sensitive to the pH of the water and cannot survive if it is too acidic. If the pH changes, they die and the animals that feed on them, such as fish, may also die.

Acid rain leaches aluminium ions out of the soil. If they reach a high concentration in the water, the gills of fish are affected. It causes the fish to suffocate.

Soot and smog

The soot particles in the air from smoke settle on buildings and plant life. They make buildings dirty and form black coatings on their outer surfaces. When soot covers leaves, it cuts down the amount of light reaching the leaf cells and slows down photosynthesis. As well as being used in industry, coal used to be the main fuel for heating homes in the United Kingdom until the 1950s. In foggy weather the smoke from the coal combined with water droplets in the fog to form smog. The water droplets absorbed the soot particles and chemicals in the smoke and made a very dense cloud at ground level, through which it was difficult to see.

Figure 5.7 The London smog of 1952

6 What property of soot particles affects photosynthesis?

7 A lake is situated near a factory that burns coal. How may the lake be affected in years to come if:

a) there is no control of emissions at the factory

b) there is no control of similar emissions worldwide?

Explain your answers.

When people inhaled air containing smog the linings of their respiratory systems became damaged. People with respiratory diseases were particularly vulnerable to smog and, in the winter of 1952, 5000 people died in London. This tragedy led to the passing of laws to help reduce air pollution.

In Los Angeles, weather conditions in May to October led to the exhaust gases from vehicles and smoke from industrial plants collecting above the city in a brown haze. Sunlight shining through this smog causes photochemical reactions to occur in it. This produces a range of chemicals including peroxyacetyl nitrate (PAN) and ozone. Both of these chemicals are harmful to plants and ozone can produce asthma attacks in the people in the city.

The Gaia Hypothesis

The Gaia Hypothesis was developed to explain how life survives on Earth. It is the idea of James Lovelock who was born in England in 1919. He studied chemistry at university then began researching in medical science. He is also a keen inventor and in the 1950s he devised scientific instruments that could detect gases occurring in very small amounts in the air. Some of these gases – such as CFCs, which destroy ozone – have been shown to seriously affect the environment. From this work he was invited to take part in planning the Viking mission to Mars.

James Lovelock was particularly interested in the atmosphere of Mars but found that the main plan was to search for life on the planet by examining the soil. He continued with his thoughts about planet atmospheres and life and examined data about the atmospheres of Mars, Venus and Earth. He found that the atmospheres of Mars and Venus both had large amounts of carbon dioxide and tiny amounts of nitrogen and oxygen. According to calculations, these amounts matched those that would be found if chemical reactions between elements and compounds in the rocks and atmospheres were the only processes that had taken place. When he examined the Earth's atmosphere he found it to have a very small amount of carbon dioxide, a huge amount of nitrogen and about one fifth of it was oxygen. He concluded that this difference was due to the presence and activities of living things on our planet.

When Lovelock studied data about the Earth's atmosphere in the past, before there was evidence of living things, he found that it used to be more similar to that of the other planets. Then, as living things developed and increased in number on the Earth, the atmosphere changed to containing a large amount of oxygen. Data about the atmosphere during the time that living things have been present shows that its composition has remained fairly constant – the amounts of the gases have stayed about the same. This led Lovelock to propose that the living things in

Figure A James Lovelock

some way regulate the composition of the atmosphere and keep it suitable for living processes. He suggested that the Earth and all its living things could be thought of as one gigantic organism, which somehow tries to keep itself alive in space – an idea that became known as the Gaia Hypothesis (Gaia was an Ancient Greek word for Earth). This conclusion was not accepted by many scientists and most thought that it was more like science fiction than science fact.

Lovelock responded to the criticisms of his hypothesis in 1983 by devising a computer simulation called Daisyworld with Andrew Watson, a scientist who also studies the atmosphere. The simulation features a Sun and an Earth-type planet, which has seeds of daisies with black petals and seeds of daisies with white petals on its surface. The black daisies have an advantage in cooler conditions, because their dark colour means they can absorb heat better. The white daisies do better in hotter conditions because they reflect more radiation, and therefore do not dehydrate so easily.

At the beginning, the Sun shines and heats up the Earth. When a certain temperature is reached, black daisy seeds germinate and grow. The black surface formed by the daisies causes the planet to absorb more heat and makes the temperature rise further. This causes the germination of white daisy seeds, which out-compete the black daisies in high-temperature conditions. As the white daisies spread across the planet, the whiteness reflects heat and the planet cools, so that black daisies start to do better again. Eventually, a 'normal' temperature is reached at which both plants can survive.

1 When James Lovelock changed from chemistry to medical science what sign of a scientist was he showing?

2 What two signs of a scientist did Lovelock show when he became an inventor?

3 What evidence did Lovelock use to conclude that living things changed the chemical composition of the atmosphere?

4 What evidence did Lovelock use to conclude that living things regulated the composition of the atmosphere in some way?

5 What further evidence did Lovelock present that suggested that living things could alter conditions on a planet?

There are equal numbers of white and black daisies.

The temperature on Daisyworld is 'normal'.

There are mostly black daisies.

The temperature on Daisyworld becomes 'high'.

There are mostly white daisies.

The temperature on Daisyworld becomes 'low'.

Figure B Daisyworld is a model that helps to explain the Gaia Hypothesis.

Both Lovelock and Watson stated that this simple model could not be directly compared to conditions on Earth but it did show how activities on the planet could regulate the conditions. Since then the model has been made more complicated with the addition of animal species such as rabbits and foxes and this has shown that more species make the regulation of the planet temperature more efficient.

There is still much scientific debate about the Gaia hypothesis but it has led to the establishment of Earth Science in which the atmosphere, the oceans, the rocky part of the land and the biosphere are studied to find out how each may affect the others. The scientists who take part in these investigations are biologists, chemists and physicists and through their work the effects of the activities of humans on the environment are being investigated.

> ### For discussion
>
> **Could the Earth be a gigantic organism in space? What signs of life does it show? Explain your answers.**
>
> **Human activities upset the composition of the atmosphere and seas. If we think of the Earth as a gigantic organism, what might we think of people as being in this model?**

Water pollution

Fresh water

Fresh water, such as streams and rivers, has been used from the earliest times to flush away wastes. Over the past few centuries many rivers of the world have been polluted by a wide range of industries including textile and paper-making plants, tanneries and metal works. People in many countries have become aware of the dangers of pollution and laws have been passed to reduce it. Ways have been found to prevent pollution occurring and to recycle some of the materials in the wastes.

Figure 5.8 This cellulose factory in Russia is causing the water to become polluted.

8 How are the lives of people who live by polluted rivers and catch fish from them put at risk?

9 The water flowing through a village had such low levels of mercury in it that it was considered safe to drink. Many of the villagers showed signs of mercury poisoning. How could this be?

The most harmful pollutants in water are PCBs (polychlorinated biphenyls) and heavy metals such as cadmium, chromium, nickel and lead. In large concentrations these metals damage many of the organs of the body and can cause cancers to develop. PCBs are used in making plastics and, along with mercury compounds, are taken in by living organisms at the beginning of food chains (page 49). They are passed up the food chain as each organism is eaten by the next one in the chain. This leads to organisms at the end of the food chain having large amounts of toxic chemicals in their bodies, which can cause permanent damage or death.

The careless use of fertilisers allows them to drain from the land into rivers and lakes and leads to the overgrowth of water plants, including algae. This is called algal bloom. When these plants die, large numbers of decomposing bacteria take in oxygen from the water. The reduction in oxygen levels in the water kills many water animals. Phosphates in detergents also cause an overgrowth in water plants, which can lead to the death of water animals in the same way.

Figure 5.9 The excessive use of fertilisers leads to algal bloom in rivers and kills fish.

Seawater

The pollutants of fresh water are washed into the sea as rivers reach the coast, where they may collect in the coastal marine life. The pollutants may cause damage to the plants and animals that live in the sea and make them unfit to be collected for human food.

Large amounts of oil are transported by tankers across the ocean every day. In the past, the tanker crew flushed out the empty oil containers with seawater to clean them. The oil that was released from the ship formed a film on the water surface, which prevented oxygen from entering the water from the air. It also reduced the amount of light that could pass through the upper waters of the sea to reach the phytoplankton (tiny floating plants) and allow them to photosynthesise.

The problem of this form of oil pollution has been reduced by adopting a 'load on top' process, where the water used to clean out the containers is allowed to settle and the oil that has been collected floats to the top. This oil is kept in the tanker and is added to the next consignment of oil that is transported.

Occasionally a tanker is wrecked at sea, or there is an oil spill at an underwater well. When this happens large amounts of oil may spill out onto the water and be washed up onto the shore. This causes catastrophic damage to the habitat, and even with the use of detergents and the physical removal of the oil the habitat may take years to recover (Figure 5.10).

10 How does oil floating on the surface of the sea affect the organisms living under it?

Figure 5.10 In April 2010, there was an explosion at the Deepwater Horizon oil well in the Gulf of Mexico, which resulted in about 780 000 m³ of crude oil flowing into the sea over the next three months. Oil spills like this can have a huge effect on marine and coastal wildlife. This photo shows how fishing boats were used to try to clear up oil floating on the sea surface.

Indicators of pollution

Some living things are very sensitive to pollution and therefore can be used as biological indicators of pollution.

Lichens are sensitive to air pollution (Figure 5.11). Where the air is very badly polluted, no lichens grow but a bright green Protoctista called *Pleurococcus* may form a coating on trees. Crusty lichens, some species of which are yellow, can grow in air where there is some pollution. Leafy lichens grow where the air has only a little pollution. Bushy lichens can grow only in unpolluted air.

Some freshwater invertebrates can be used to estimate the amount of pollution in streams and rivers (Figure 5.12). If the water is very badly polluted there is no freshwater life, but if the water is quite badly polluted rat-tailed maggots may be present. Bloodworms can live in less badly polluted water and freshwater shrimps can live in water that has only small amounts of pollution. Stonefly nymphs can live only in unpolluted water.

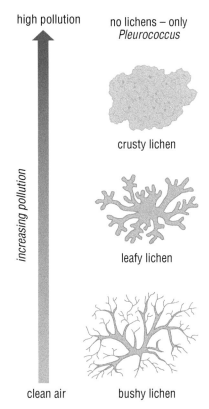

high pollution

no lichens – only *Pleurococcus*

crusty lichen

increasing pollution

leafy lichen

bushy lichen

clean air

Figure 5.11 Lichens can indicate air quality.

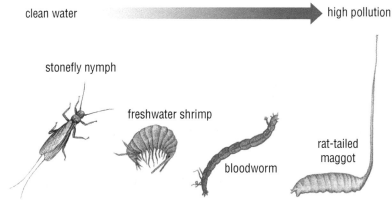

clean water   high pollution

stonefly nymph

freshwater shrimp

bloodworm

rat-tailed maggot

Figure 5.12 Some freshwater invertebrates

11 How polluted is the habitat if:
   a) the trees have *Pleurococcus* and crusty yellow lichens on them
   b) bushy, leafy and crusty lichens are found in a habitat
   c) freshwater shrimps and bloodworms are found in a stream?

12 Four places were examined in succession along a river and the animals found there were recorded. Here are the results.
   Station A: Stonefly nymphs, mayfly nymphs, freshwater shrimps, caddisfly larvae
   Station B: Rat-tailed maggots, sludge worms
   Station C: Sludge worms, bloodworms, waterlice
   Station D: Freshwater shrimps, waterlice
   What do the results show? Explain your answer.

Intensive farming

As the human population has continued to grow (page 60) ways have been found to increase food production by intensive farming. A feature of intensive farming is the use of fertilisers and pesticides.

Fertilisers

When plants are grown for food they are eventually harvested and taken out of the soil. This means that the minerals they have taken from the soil go with them to market, and there are therefore fewer minerals left in the soil for the next crop. **Fertilisers** are added to the soil to replace the minerals that have been taken away in the crop. They are added in quantities that will make sure that the plants grow as healthily as possible and produce a large crop. The amount of fertiliser added to the soil has to be carefully calculated. If too little is added, the amount of food produced by the crop (called the yield) will be small. If too much is added, the plants will not use all the minerals and they may be washed into streams and rivers and cause pollution.

There are two kinds of fertiliser. Inorganic fertilisers such as ammonium nitrate are manufactured chemical compounds. Organic fertilisers are made from the wastes of farm animals (manure) and humans (sewage sludge).

Figure 5.13 This farmer is spreading fertiliser on the field to ensure that the crop will have enough essential minerals to give a good yield.

Inorganic fertilisers can give crops an almost instant supply of minerals, as the minerals dissolve in the soil water as soon as they reach it, and can be taken up by the roots straight away. The minerals in manure are released more slowly, as decomposers (page 55) in the soil break it down.

Inorganic fertilisers are light in weight and so can be spread from aeroplanes and helicopters flying over the crops. This means that they can be applied to the crop at any time without damaging it. Manure is too heavy to be spread in this way and must be spread from a trailer attached to a tractor. If the manure was spread while the crop was growing the tractor and trailer would damage the crop, so the manure must instead be spread on the soil before the crop is sown. This also allows some time for the manure to release minerals into the soil.

13 What kind of fertiliser gives you greater control over providing minerals for a crop? Explain your answer.

Fertilisers and soil structure

A good soil has particles of rock bound together with **humus** (decomposed plants and animals) to form lumps called crumbs. The crumbs do not interlock but settle on each other loosely, with air spaces between them. The air provides a source of oxygen for the plant roots and for organisms living in the soil. The spaces also allow the roots to grow easily through the soil. The humus acts like a sponge and holds onto some of the water that passes through the soil. The plant roots draw on the water stored in the humus.

In time humus rots away, and in natural habitats it is replaced by the decaying bodies of other organisms. When manure is added to the soil it adds humus and this helps to bind the rock particles together and keep the soil crumbs large. If the soil receives only inorganic fertilisers the humus is gradually lost and the soil crumbs break down. The rocky fragments that remain form a dust, which can be easily blown away by the wind.

14 If a farmer uses only inorganic fertilisers how may the soil organisms be affected?

Pests and pesticides

Fertilisers are used to make the crop yield as high as possible. The crop plants may, however, be affected by other organisms, which either compete with them for resources, or feed on them. These organisms are known as pests, and chemicals called **pesticides** have been developed to kill them. There are three kinds of pesticide – herbicides, fungicides and insecticides.

Herbicides

When a crop is sown the seeds are planted so that the plants will grow a certain distance apart from each other. This distance allows each plant to receive all the sunlight, water and minerals that it needs to grow healthily and produce a large yield.

A weed is a plant growing in the wrong place. For example, poppies may be grown in a flowerbed to make a garden look attractive, but if they grow in a field of wheat they are weeds because they should not be growing there. Weeds grow in the spaces between the crop plants and compete with them for sunlight, water and minerals. This means that the crop plants may receive less sunlight because the weeds shade them, and receive less water and minerals because the weed plants take in some for their own growth. Weeds can also be infested with microorganisms that can cause disease in the crop plant. For example, cereals may be attacked by fungi that live on grass plants growing as weeds in the crop.

Figure 5.14 Poppies growing as weeds in a field of wheat

Weeds are killed by **herbicides**. There are two kinds of herbicide – non-selective and selective. A non-selective herbicide kills any plant. It can be used to clear areas of all plant life so that crops can be grown in the soil later. It must not be used when a crop is growing as it will kill the crop plants too. A selective herbicide kills only certain plants – the weeds – and leaves the crop plants unharmed. It can be used when the crop is growing.

Herbicides may be sprayed onto crops from the air. Some of the herbicide may drift away from the field and into surrounding natural habitats. When this happens, the herbicide can kill plants there. Many wild flowers have been destroyed in this way.

15 How might the use of herbicides on a farm affect the honey production of a local bee keeper? Explain your answer.

Fungicides

A **fungicide** is a substance that kills fungi. Fungal spores may be in the soil of a crop field, floating in the air, or on the seeds before they are sown. Seeds may be coated with fungicides to protect them when they germinate. They are also applied to the soil to prevent fungi attacking roots, and are sprayed on crop plants to give a protective coat against fungal spores in the air.

Insecticides

When a large number of plants of the same kind are grown together, they can provide a huge feeding area for insects. Large populations of insects can build up on the plants and cause great damage. Perhaps the worst insect pest of all is the locust. A swarm of locusts can eat entire fields of plants.

Figure 5.15 Swarms of locusts like this one threaten crops in many warmer parts of the world.

Insecticides are used to kill these insect pests. There are two kinds of insecticide – narrow-spectrum and broad-spectrum. Narrow-spectrum insecticides are designed to kill only certain kinds of insects, and leave others unharmed. Broad-spectrum insecticides, in contrast, kill a wide range of insects – not only those that feed on the crop, but also predatory insects that may prey on them. If insecticides drift away from the fields after spraying, they can kill other insects in their natural habitats. They can also be passed along the food chain.

16 You have a field of weeds and need to plant a crop in it. How could you use pesticides to prepare the ground for your crop and protect it while it is growing?

A poison in the food chain

In 1935, a Swiss chemist called Paul Müller (1899–1965) set up a research programme to find a substance that would kill insects but would not harm other animals. Insects were his target because some species are plant pests and devastate farm crops, and others carry microorganisms that cause disease in humans. The substance also had to be cheap to make and not have an unpleasant smell. In 1939 he tried a chemical called dichlorodiphenyltrichloroethane (DDT), which was first made in 1873. DDT seemed to meet all the requirements and soon it was being used worldwide.

In time, some animals at the end of the food chains (the top carnivores) in the habitats where DDT had been sprayed to kill insects were found dead. The concentration of the DDT applied to the insects was much too weak to kill the top carnivores directly so investigations into the food chains had to be made.

In Clear Lake, California, DDT had been sprayed onto the water to kill gnat larvae. The concentration of DDT in the water was only 0.015 parts per million (ppm), but the concentration in the dead bodies of fish-eating water birds called western grebes was 1600 ppm. When planktonic organisms in the water were examined, their bodies contained 5 ppm and the small fish that fed on them contained 10 ppm.

Figure 5.16 Western grebes suffered at Clear Lake in California, USA, because DDT sprayed to kill insect larvae accumulated in their bodies at high concentrations.

17 Construct the food chain investigated in Clear Lake.

18 Why did the grebes die?

It was discovered that DDT did not break down in the environment but was taken into living tissue and stayed there. As the plankton in the lake were eaten by the fish the DDT was taken into the fishes' bodies and built up after every meal. The small fish were eaten by larger fish in which the DDT formed higher concentrations still. The grebes ate the large fish and with every meal increased the amount of DDT in their bodies until it killed them.

In Britain, the peregrine falcon is a top carnivore in a food chain in moorland habitats, although it visits other habitats outside the breeding season. The concentration of DDT in bodies of female falcons caused them to lay eggs with weak shells. When parents incubated the eggs their weight broke the shells and the embryos died.

Biological pest control

Biological pest control is the use of one animal species to control the numbers of a pest. In 300 CE, the Chinese began using biological pest control to protect their crops of mandarin oranges. These fruits are attacked by many insects, but it was discovered that a certain kind of large ant did not eat the crop but fed on the harmful insects. The ants were collected by filling sheep bladders with fat and putting them next to ants' nests. When the bladders were teeming with ants they were collected and placed in the orange trees. Bamboo poles were placed from one tree to the next in the orange grove so that the ants could move freely over all the trees, and eat the pest insects that would otherwise damage the fruit. This form of biological control is still used today.

The biological control of other pests did not begin in other parts of the world until much more recent times. In Australia, the prickly pear was introduced as an ornamental garden plant but it soon started growing wild and became a weed. A moth that has a larva that feeds on the prickly pear was released among prickly pears, and in time their numbers were greatly reduced.

Figure 5.17 Ladybirds are used to control the numbers of aphids, which are a common plant pest.

For discussion

There are larger pests than insects. For example, mice nibble roots and rabbits eat shoots. What form of biological control could be used to keep down the numbers of mice and rabbits?

Should only natural predators or primary consumers be used for biological control or should new kinds of animal be bred to control pests? Explain your answer.

The perfect environment for growth

When plants are grown for food, every attempt is made to ensure that the crop yield is as large as possible.

Applying fertilisers helps crop growth, and the use of pesticides keeps other organisms from damaging the crop. There are, however, other factors that affect growth. They are the factors that affect photosynthesis – light, temperature and the amount of carbon dioxide in the air

(Chapter 1). Figures 5.18 to 5.20 show the results of experiments investigating the effects of light intensity, temperature and the concentration of carbon dioxide in the air on the rate of photosynthesis.

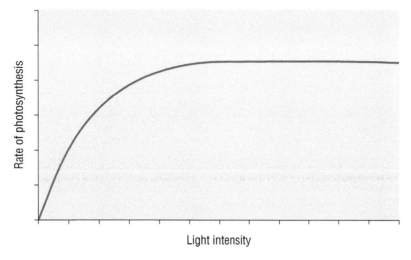

Figure 5.18 The effect of light intensity on the rate of photosynthesis

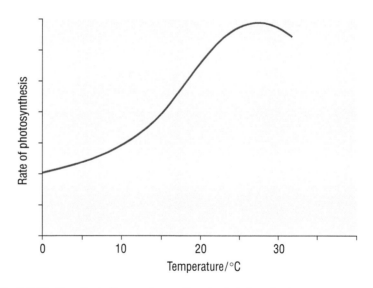

Figure 5.19 The effect of temperature on the rate of photosynthesis

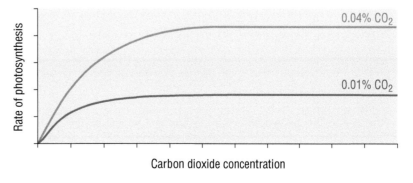

Figure 5.20 The effect of carbon dioxide concentration on the rate of photosynthesis

Combining the three factors

The results of the experiments show that the amount of light, the temperature and the amount of carbon dioxide in the air all affect the rate of photosynthesis. This implies that the rate of photosynthesis can be increased to a maximum by carefully regulating these three factors in the environment of the plant. These factors cannot be controlled in an open field but they can be controlled by using a glasshouse.

Figure 5.21 In this huge glasshouse the perfect environment is being created for growing cucumber plants.

◆ SUMMARY ◆

◆ From the earliest civilisations, human activity has caused some pollution but the problem greatly increased with the Industrial Revolution (*see page 65*).

◆ Human activity has caused holes in the ozone layer (*see page 68*).

◆ Human activity has caused a rise in the amount of carbon dioxide in the atmosphere, which may be contributing to global warming (*see page 69*).

◆ Human activity has created acid rain in some places (*see page 69*).

◆ Soot and smog have been caused by human activity (*see page 70*).

◆ The creative thinking of James Lovelock has produced a new way of looking at the Earth (*see page 71*).

◆ Human activity has caused water pollution (*see page 73*).

◆ Intensive farming requires the use of fertilisers and pesticides (*see page 77*).

◆ Some pests are controlled by biological pest control (*see page 82*).

◆ Glasshouses can be used to create the perfect environment for growing crops (*see page 84*).

End of chapter questions

Figure 5.22 shows the position of two coastal towns. A and B are two towns that have a fishing industry. Due to over-fishing, the industry has declined. There are large numbers of people now unemployed in both towns and many are thinking of moving or travelling to the cities to find work. It is proposed to build an oil refinery near one of the towns. This will bring employment for the people in the form of building and running the refinery, and in the factories that may be set up to use its products. Land will be needed for the refinery and for the port where the oil tankers will dock. Land will also be needed for factories and perhaps housing estates if more people come to live in the town.

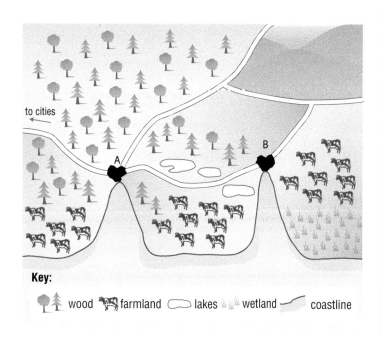

Key:

🌳 wood 🐄 farmland ⬭ lakes ⸓⸓ wetland ⌒ coastline

Figure 5.22 Map showing the positions of towns A and B

1 What else may land be needed for if more people come to the town?
2 What habitats may be affected by the building of the oil refinery? Explain your answers.

For discussion

What are the advantages and disadvantages of choosing to build the refinery near town A or town B?

What are the major issues involved in deciding where the refinery is to be built?

How would you balance these issues to decide which town is best suited for the refinery and the port?

For discussion

Imagine that you are going to live on an island with about two dozen other people. The island has four seasons, with a warm summer and a cool winter. It has a hill near its centre, which provides some shelter from the cold prevailing wind. Much of the low land on the sheltered side of the hill is woodland. There is a small population of fish in the sea around the island, which can provide a little food. Work out a plan for providing food for yourself and the rest of the group for a stay of two years on the island. You may take any plants, animals, pesticides, fertilisers and building materials that you wish, and you may begin your stay on the island in any season.

6 Classification and variation

- The main groups of living things
- Keys
- Inherited characteristics
- How cells reproduce
- Chromosomes and genes
- Chromosomes and gametes
- Genes and variation
- Selective breeding
- Natural selection
- Charles Darwin

When you come across an unfamiliar animal in a habitat survey you might look at it to see if it reminds you of anything else. If it has six legs you might compare it with an insect that you know. If it is a bird you may compare its plumage with a bird you are familiar with. This comparing, looking for similarities and differences, led scientists to develop keys and to classify all the living things on Earth into groups.

The main groups of living things

Microorganisms

Table 6.1 Microorganisms can be divided into four groups.

Group	Features	Examples
Fungi kingdom	have spores, and feeding threads called hyphae	yeast, mushroom
Monera kingdom	one cell body without a nucleus	bacteria, blue-green algae
Protoctista kingdom	one cell body with a nucleus	amoeba, ciliates
viruses	no cell structure	influenza virus, human immunodeficiency virus (HIV)

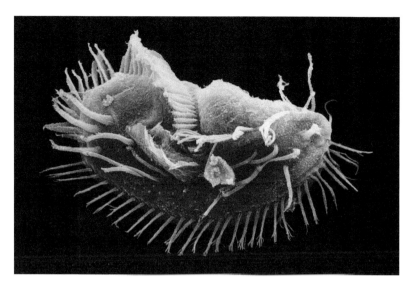

Figure 6.1 The cell wall of a ciliate protozoan, from the Protoctista kingdom, is covered in tiny hairs called cilia, which it uses like oars to row itself through the water.

The plant kingdom

Table 6.2 The plant kingdom is made up of different groups.

Group	Features	Examples
algae	do not have roots, stems or leaves	green slime, seaweeds
liverworts and mosses	do not have true roots, have spores	sphagnum moss, common or umbrella liverwort
ferns	roots, stems, fronds and spores	fiddle head ferns
conifers	roots, woody stem, cones, seeds	fir, pine
flowering plants	roots, stem, leaves, flowers, seeds some are woody and others non-woody	daisy, mango, mint, oak, pineapple, fig, chestnut, dandelion

Figure 6.2 Passion flowers have a distinctive arrangement of stigmas, stamens, petals and sepals to attract a range of pollinators including bees, wasps, hummingbirds and bats.

The animal kingdom

The animal kingdom is divided into two groups – the invertebrates (animals without a backbone) and vertebrates (animals with a backbone).

Invertebrates

Table 6.3 There are many types of invertebrate.

Group	Features	Examples
jellyfish	soft body, tentacles at one end round the mouth	box jellyfish, sea anemone, coral (make hard casing)
flatworms	very thin, flat body	flukes, tapeworms
annelid worms	cylindrical bodies divided into rings	earthworm, leech, ragworm, fanworm
nematode worms	small cylindrical bodies not divided into rings	hookworm, pinworm
arthropods	jointed legs	insects, spiders, scorpions, crustaceans (shrimps and crabs), centipedes, millipedes
molluscs	soft body with shell (which might be inside)	snails, slugs, octopus, mussels, squid, scallops, limpets
echinoderms	spiny skin, arms (starfish) or globe shaped (sea urchins)	brittle star, sea cucumber, sea lily

Figure 6.3 A starfish is a type of invertebrate called an echinoderm. It has no head, but has a simple brain in the form of a ring around its mouth. Each arm has large numbers of tiny feet with suckers to help it move along the sea floor and pull open mussels' shells to feed on them.

Vertebrates

Table 6.4 There are five groups of vertebrates.

Group	Features	Examples
fish	scales, fins	barracuda, shark
amphibians	smooth or warty skin, tadpole stage	frogs, toads, salamanders
reptiles	skin covered in dry scales	snakes, turtles, lizards
birds	feathers, beaks	hummingbird, eagle
mammals	fur, give milk to young	mouse, lion

1 Identify one adaptation in each of the living things shown in Figures 6.1 to 6.4 that helps it to survive in its habitat.

Figure 6.4 A mole is a mammal. It has an extra digit in each hand to make large digging structures for burrowing through soil. Moles are found in Asia, Europe and parts of North America.

Keys

2 In a key about fish, the species are identified by the length of their bodies and by their body mass. Is this key reliable? Explain your answer.

There are two kinds of **key** – spider keys (whose name refers to the way the key spreads out like the legs of a spider) and numbered keys, which take you step by step to the identification of the specimen you have found. Keys are made by considering features in a species that do not vary. Features that might change due to growth, for example, are not used.

Spider key

On each 'leg' of the spider is a feature that is possessed by the living thing named below it. An example is shown in Figure 6.5. A spider key is read by starting at the top in the centre and reading the features down the legs until the specimen is identified.

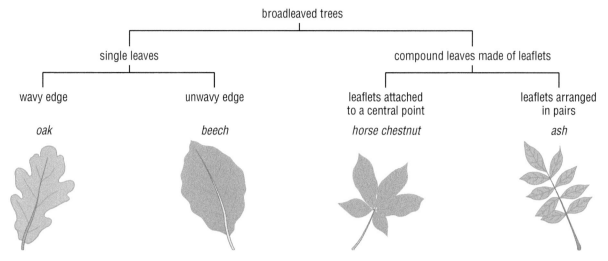

Figure 6.5 Spider key of leaves from broadleaved trees

Numbered key

You work through a numbered key by reading each pair of statements and matching the description of one of them to the features you see on the specimen you are trying to identify. At the end of each statement there is an instruction to move to another pair of statements or to the name of a living thing.

Here is a simple numbered key. It can be used to identify molluscs that live in freshwater habitats such as rivers, lakes and ponds.

1 a) Single shell	see **2**
b) Two shells	see **6**
2 a) Snail with a plate that closes the shell mouth	***Bithynia***
b) Snail without a plate that closes the shell mouth	see **3**
3 a) Snail without a twisted shell	**freshwater limpet**
b) Snail with a twisted shell	see **4**
4 a) Shell in a coil	**ramshorn snail**
b) Shell without a coil	see **5**
5 a) Snail with triangular tentacles	**pond snail**
b) Snail with long, thin tentacles	**bladder snail**
6 a) Animal has threads attaching it to a surface	**zebra mussel**
b) Animal does not have threads attaching it to a surface	see **7**
7 a) Shell larger than 25 mm	**freshwater mussel**
b) Shell smaller than 25 mm	**pea mussel**

3 Make a spider key for the four animals in Figure 6.6. Look carefully at the animals. Choose a feature they have all got in common to put at the top of the key.

Figure 6.6 Molluscs and annelid worms

4 a) Identify the molluscs **a–f** in Figure 6.7 using the numbered key on page 90. In each case, write down the number of each statement you used to make the identification. For example, specimen **a** is identified by following statements 1a, 2b, 3a. It is a freshwater limpet.
b) Why should another feature in addition to size be added to the statements in part 7 of the key?

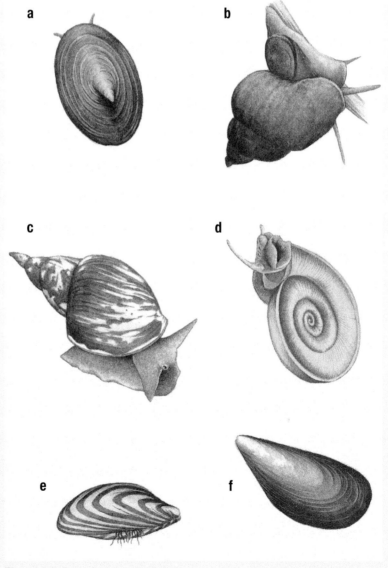

Figure 6.7 Some freshwater molluscs

5 Make up a numbered key to identify the arthropods in Figure 6.8. Begin by separating the butterfly, which has six legs, from the others.

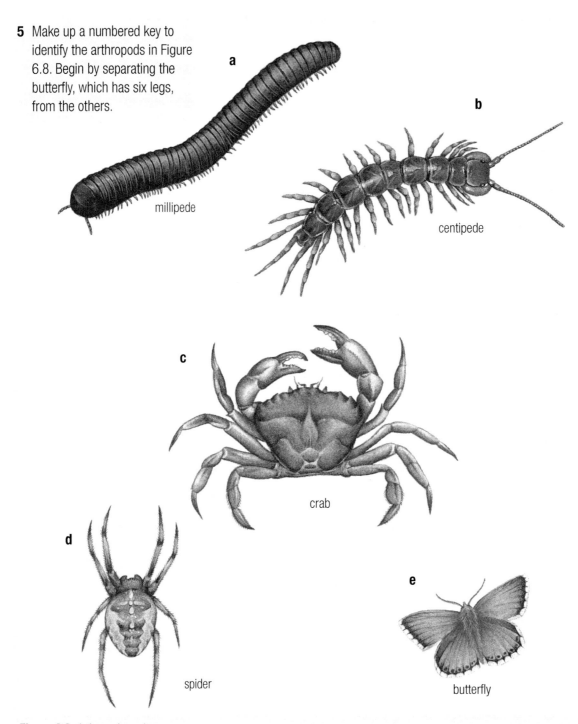

a
millipede

b
centipede

c
crab

d
spider

e
butterfly

Figure 6.8 Arthropod specimens

6 Look at the parts of the insect's body in Figure 6.9a. Then look at how the parts vary in the six insect specimens in Figure 6.9b. Invent a key to distinguish between these six specimens.

7 Look at the spider key for the leaves, and the numbered key for freshwater molluscs (page 90), and answer the following questions.

   a) Which key identifies the larger number of living things?

   b) If both keys featured the same number of living things, which key would need the larger amount of space?

   c) Give an advantage of a numbered key.

   d) Give an advantage of a spider key.

   e) Which is the better one to use in a poster? Explain.

   f) Which is the better one to use in a pocket book for fieldwork? Explain.

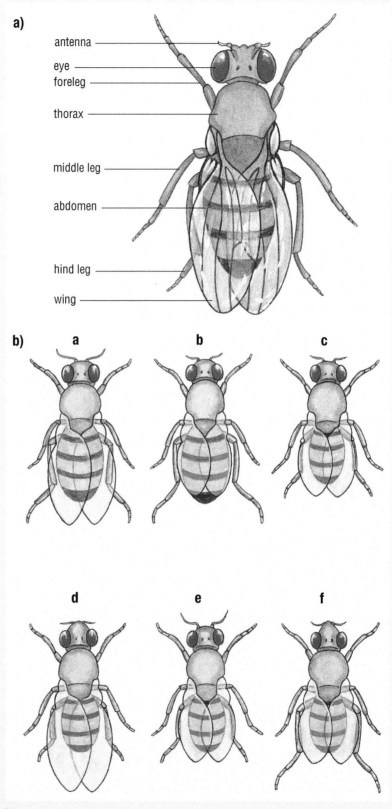

Figure 6.9 Variation in flies

Inherited characteristics

The features of living things that we have looked at when considering adaptations and making keys are also known as their **characteristics**. When you look at different generations of the same species you see that they have a large number of the same characteristics. It appears that the characteristics are passed on from one generation to the next. The latest generation has inherited its characteristics from the generation before. Once scientists grasped this idea, investigations into inherited characteristics began.

Gregor Mendel and Hugo de Vries

Gregor Mendel (1822–1884) was an Austrian monk who studied mathematics and natural history. He set up experiments to investigate how features in one generation of pea plants were passed on to the next.

Pea flowers self-pollinate (page 22). When Mendel wished to control the way the flowers pollinated he cut off the anthers of one flower, collected pollen from another flower and brushed it on to the stigma of the first. He completed his task by tying a muslin bag around the first flower to prevent any other pollen from reaching it.

Mendel performed thousands of experiments and used his mathematical knowledge to set out his results and to look for patterns in the way that the plant features were inherited. He suggested that each feature was controlled by an inherited factor. He also suggested that each factor had two sets of instructions and that parents pass on one set of instructions each to their offspring. Many years later it was discovered that Mendel's 'factors' were genes (pages 95–98).

Mendel's work was published by a natural history society but its importance was not realised until 16 years after his death. At that time, Hugo de Vries (1848–1935), a Dutch botanist, had been studying how plants pass on their characteristics from generation to generation. He was checking through published reports of experiments when he discovered Mendel's work. He found that his own work supported Mendel's and after studying Charles Darwin's hypothesis, called Pangenesis, to explain inheritance he went on to suggest that information about the characteristics to be inherited were in the form of small particles called pangenes. Darwin's Pangenesis hypothesis was shown to be incorrect but de Vries' idea about pangenes continued to be useful to scientists studying inheritance and a Danish biologist called Wilhelm Johannsen (1857–1927) made the word shorter and called the inherited particles 'genes'.

Figure A Gregor Mendel

1 Why did Mendel cut out the anthers of some flowers?
2 Why did Mendel tie a muslin bag around the flowers in his experiments?

3 What evidence provided Mendel with the idea of using pollen studies to investigate inheritance?
4 What is the value of performing a large number of experiments?
5 How did Mendel's mathematical knowledge help him in considering the evidence of his investigations?

6 Which scientific enquiry skills was de Vries applying when he discovered Mendel's work?
7 What evidence stimulated de Vries' creative thoughts about inherited particles?

Cells and reproduction

As living things are made from cells, scientists began looking at cells for evidence that they may be involved in reproduction. We have seen from slides of tissues that cells have a cell membrane, cytoplasm and a **nucleus**. When scientists examined living cells under the microscope, they saw that changes took place in the nucleus that are linked to reproduction. Just before a cell divides long strands of material appear in the nucleus. These strands are called **chromosomes**.

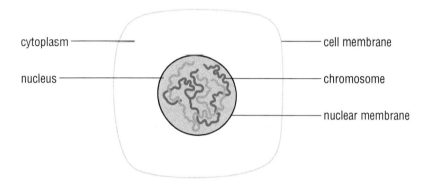

cytoplasm ———————————— cell membrane

nucleus ———— chromosome

———— nuclear membrane

Figure 6.10 Chromosomes in the nucleus of a cell

In the time before cell division begins, when the chromosomes are not visible under the microscope, each one has made a copy of itself, and the chromosome and its copy lie together as two threads. During cell division, as the nucleus divides, each pair of threads separates. Each thread then enters one of the two nuclei that are forming in the new cells. Once each cell is formed, every thread becomes a chromosome. This means that the nuclei of the new cells have the same number of chromosomes as the original one. The chromosomes in the nuclei of the new cells then seem to disappear into the nuclear material but become visible under the microscope once more when the cells are about to divide.

Chromosomes and genes

Extensive studies on chromosomes have shown that they are arranged in pairs in the nucleus and individuals in each species have a certain number of chromosomes in their cell nuclei. For example, humans have 46 chromosomes arranged in 23 pairs in every nucleus, and a fruit fly has eight chromosomes arranged in four pairs.

The chromosomes are really threads of chemical messages. These messages are strung along each

8 Draw a sequence of pictures about a cell with only one chromosome from the time before the chromosome becomes visible in the cell until two new cells are produced and the chromosomes have disappeared into their nuclear material.

chromosome like the carriages in a very long train. Each message is called a **gene**. The genes provide all the information for how the cell grows, develops and behaves and how the body grows and develops too. Each pair of chromosomes has pairs of genes that carry information for a particular characteristic such as eye colour or hair colour. These genes are situated at the same point on each chromosome, as Figure 6.11 shows.

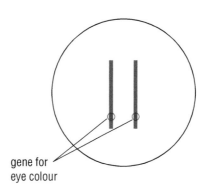

gene for
eye colour

Figure 6.11 Two genes on a pair of chromosomes

Chromosomes and gametes

Once the link between chromosomes and the reproduction of body cells had been worked out, scientists looked for a link between chromosomes and the cells involved in reproduction of whole individuals. These cells are called the sex cells or gametes. In animals, the male gamete is the sperm and the female gamete is the egg or ovum. In plants, the male gamete is a cell enclosed in a pollen grain and the female gamete, called the egg cell, is in the ovule. During fertilisation the nuclei of the male and female gametes join together and a cell called a zygote is formed. This then grows into a new individual.

If the gametes were produced in the same way as ordinary body cells they would have the same number of chromosomes as body cells. This would create a problem at fertilisation as the zygote would have twice as many chromosomes as the parents. If the zygotes from this generation formed individuals that bred, the new zygotes would have four times as many chromosomes as the grandparents. After a few more generations the zygotes would be so packed with chromosomes that they would die. This does not happen because there is a special type of cell division that takes place in the reproductive organs, which produces gametes with half the number

9 The cells of a pineapple plant have 50 chromosomes in their nuclei.

a) How many chromosomes are there in the male nucleus in a pineapple pollen grain?

b) How many are there in the female nucleus in the ovule of a pineapple flower.

c) How many chromosomes are in the cells of a seedling?

of chromosomes in body cells. In this cell division the chromosomes do not make copies of themselves but the pairs separate and move into the newly forming cells – the gametes. For example, the cells in the human reproductive organs pass on 23 chromosomes into each of the gametes they produce.

At fertilisation the chromosomes from the gametes pair up in the zygote and the production of body cells begins again, as Figure 6.12 shows.

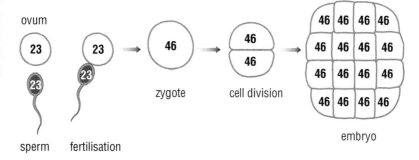

Figure 6.12 How chromosomes pair up at fertilisation and produce a new individual

Genes and variation

During the formation of the gametes, parts of the chromosomes swap portions. This swapping leads to a mixing up of the genes so an exact copy of the parent's genetic code is not passed on. When a zygote is formed after fertilisation the nucleus contains all the genes needed to make the new individual. As there has been some mixing of the genes from both parents, the new individual develops a slightly different combination of features from their parents, which leads to **variation** in the species.

How genes work together

It is easy to imagine the genes of a single-celled organism such as a ciliate in Figure 6.1 carrying out their tasks to keep the one cell alive, but what happens in a multicellular organism like a human? It starts off as a zygote but soon forms an embryo as Figure 6.12 shows. Its chromosomes contain all the genes to make the body and keep it alive. How do these genes work together? The answer is that as the cells divide and form the embryo and foetus, all the genes in their nuclei are not used or switched on at once. When it is time for an organ to develop, the genes that control its development switch on and the cells that are produced make the organ.

Other genes also present on the chromosomes but controlling the development of other organs do not switch on in these cells. For example in the production of the windpipe or trachea the genes of some cells will switch on to make ciliated epithelial cells.

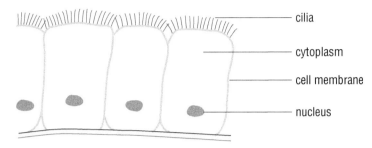

Figure 6.13 The nuclei of these cells still contain other genes, such as for eye colour, but they remain switched off forever. When the iris of the eye is forming, genes in the iris-making cells switch on while genes for making cilia remain permanently switched off.

Variation in a family

Figure 6.14 shows a family. The father has genes for black hair, curly hair, blue eyes, attached ear lobes (the lower part of the ear is joined straight down on to the side of the head) and for the absence of freckles. The mother has genes for red hair, straight hair, brown eyes, free ear lobes, and freckles.

10 What features have
   a) Alberto
   b) Benita
   c) Carlos
   d) Dorita
inherited from their mother and from their father?

Figure 6.14 Variation in a family

You can see how the characteristics controlled by the genes can vary between one generation and the next (parents and children) and among individuals of the same generation (the children) by answering question **10**.

DNA

Genes are made from a substance called deoxyribonucleic acid which is usually shortened to DNA. The first work on investigating the chemicals in cell nuclei was carried out in 1869 by Johann Friedrich Miescher (1844–1895). He used the white cells in pus and the substance he discovered was called nuclein. Over the next 84 years generations of scientists made investigations on this substance. Rosalind Franklin (1920–1958) studied the structure of molecules by firing X-rays at them. In 1951 she investigated DNA in this way and her results suggested to her that it could be made of two coiled strands, but she was not sure.

In 1953 James Watson and Francis Crick, using some of Franklin's results to help them, worked out that DNA is made from long strands of chemicals that are coiled together to make a structure called a double helix. The chemicals are arranged in a sequence that acts as a code. The code provides the cell with instructions on how to make the other chemicals that it needs to stay alive and develop properly.

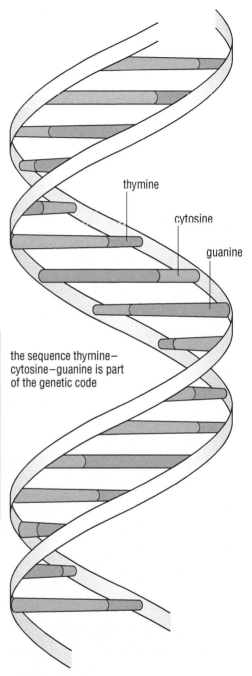

thymine

cytosine

guanine

the sequence thymine–cytosine–guanine is part of the genetic code

Figure D Basic structure of DNA

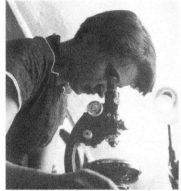

Figure B Rosalind Franklin's evidence was vital in the discovery of the structure of DNA.

Figure C James Watson and Francis Crick used evidence from several sources to work out the structure of the DNA molecule.

Barbara McClintock (1902–1992) was a biologist who studied maize – the plant that provides the food known as sweetcorn. While she was still a student she worked out a way of relating the different chromosomes in the nucleus to the features of the plant. Later, in the 1940s, she discovered that the genes on a chromosome could change position. They became known as 'jumping genes'. This discovery did not fit in with the way genes were thought to act and her work was not accepted by other scientists. But in the 1970s, during investigations by scientists on the DNA molecule, it was found that parts of the DNA broke off and moved to other parts of the chromosome. McClintock's work was proved to be correct and in 1983 she received the Nobel Prize for Physiology or Medicine.

In 1990 the Human Genome Project was begun to identify up to 25 genes on human chromosomes. This was an international project with scientists from many countries around the world working together to make it possible to see all the genes that are required to form the human body. The project was completed in 2003 and the data it has produced is being used in investigations in medical research to cure certain diseases, in the study of microorganisms that cause disease and decompose wastes, in the study of chemicals and radiation that is harmful to us and in the study of forensic science.

Figure E Barbara McClintock with the trophy for the Albert Lasker Award for Basic Medical Research she won in 1981.

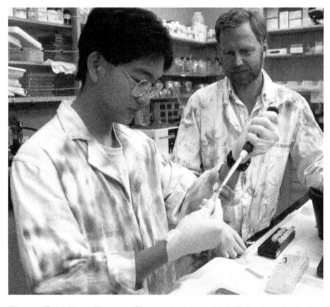

Figure F Li Ho (left) was a Chinese American High School student who worked on the Human Genome project with Dr Greg Lennin, a senior biomedical researcher.

As each person's DNA is unique it can be used for identification purposes. A person's DNA profile (sometimes called a DNA fingerprint) can be made from cells in the saliva or the blood. The DNA is chopped up by enzymes and its pieces are separated into a gel in a process like chromatography. (Remember that chromatography is the process used to separate colours in an ink by putting a drop of ink onto a paper and allowing water to soak through it.) The pattern of the pieces looks like a bar code on an item of goods. Closely related people have more similar profiles than those who are not related. Figure 5 on page 4 shows a DNA profile, with the technique's inventor, Alec Jeffreys.

8 What preliminary work did Rosalind Franklin do before she investigated DNA?

9 How did Franklin's work help Watson and Crick?

For discussion

How firmly should scientists hold their views?

For discussion

How could DNA be used to investigate a crime?

Selection

Selective breeding

For thousands of years people have been breeding animals and plants for special purposes. Most plants were originally bred to produce more food, but later plants were also bred for decoration. Animals were originally bred for domestication, then for food production or to pull carts.

A **selective breeding** programme involves selecting organisms with the desired features and breeding them together.

The variation in the offspring is examined and those with the desired feature are selected for further breeding. For example, the 'wild' form of wheat makes few grains at the top of its stalk. Individuals that produce the most grains are selected for breeding together. When their offspring are produced they are examined and the highest grain producers are selected and bred together.

By following this programme wheat plants producing large numbers of grains have been developed.

Figure 6.15 'Wild' wheat (left) and modern wheat (right)

In some breeding programmes a number of features are selected and brought together. The large number of different breeds of dog have been developed in this way.

11 All the different breeds of dog have been developed from the wolf by selective breeding. What features do you think have been selected to produce a greyhound? Give a reason for each feature you mention.

Figure 6.16 A wolf and a greyhound

Mutations

Hugo de Vries, the Dutch botanist featured with Gregor Mendel earlier in the chapter also studied the evening primrose, a plant that had been newly introduced into Holland, and found that the plants occasionally produced a new variety that was quite different from the others. This new variety was caused by mutation. Mutations had been seen in herds of livestock before by farmers, where the odd animal was sometimes called a 'sport', but de Vries was the first person to introduce the idea of mutations into scientific studies.

If a mutation survived and reproduced it could pass on its new feature to its offspring who in turn could pass it on until a new variety was established.

12 Short-legged sheep are a mutation. Could money be saved on fencing if you farmed them? Explain your answer.

13 Mutation means change. Why is it a good word to describe a new variety of a species?

Figure 6.17 Short-legged sheep mutation

Genetically modified crops

In the past selective breeding was the only way to improve crops and farm animals to provide more food. In the early 1990s the first genetically modified (GM) crop was developed. The crop plant was the tomato and it was treated so that one of its genes, which makes the fruit soften, was deactivated or switched off. The idea behind this experiment was that the tomato crop could be stored for longer. The quality of the crop was not as good as expected but work still continues on making tomato plants that are resistant to diseases.

Since the early experiment work has continued on developing genetically modified crops using genes from other organisms, such as wild relatives or from bacteria.

The aims behind developing GM foods are to help the new crop to grow better and produce a higher yield or help the food to be processed more cheaply. For example, some tomato plants were modified so that they produced tomatoes that made a thick tomato purée more cheaply than other kinds of tomatoes.

Genetically modified organisms are new organisms. Scientists disagree on their use because if they enter a natural habitat they may breed with the organisms already there to produce a more unusual organism, which may have a negative environmental impact.

Clones

A **clone** is an organism that is an exact copy of its parent. Clones can occur naturally when some organisms reproduce. *Amoeba*, a Protoctist, produces a clone by simply dividing in two.

14 How is genetic modification different from selective breeding?

15 Should plants such as sugar cane be genetically modified to provide fuel to replace oil? Should plants such as maize be genetically modified so they produce their own insecticides? Explain your answers.

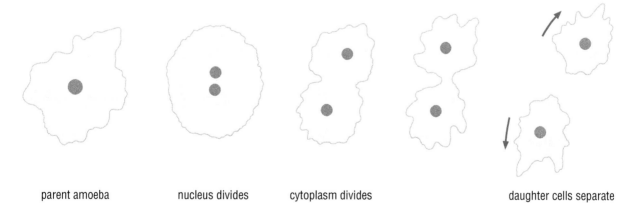

parent amoeba   nucleus divides   cytoplasm divides   daughter cells separate

Figure 6.18 Stages in the division of an amoeba

Hydra, a small animal related to sea anemones, which lives in ponds and ditches, grows a bud which detaches itself and becomes a copy of its parent.

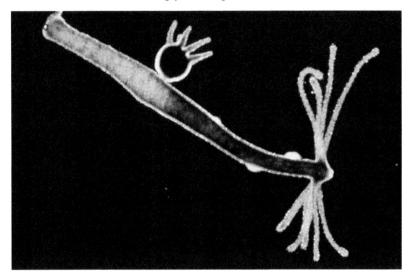

Figure 6.19 *Hydra*

Some plants can also form clones. The spider plant is a familiar houseplant which grows small plants on side shoots (Figure 6.20). The small plants can become detached and live on their own.

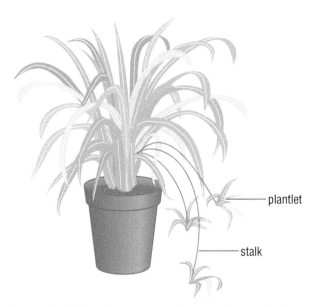

Figure 6.20 The spider plant grows in many moist woodlands in the warmer regions of the world.

A simple way of cloning a plant is to take a cutting from it and grow the cutting in compost to make a new plant.

A technique for cloning animals has been developed using nuclei and cells. The nucleus is removed from an egg cell and is replaced by the nucleus of a normal body cell from the animal you wish to clone. The egg is then allowed to develop normally and an exact copy of the animal is produced. Dolly the sheep was the first successfully reared clone at the end of the 20th century. Since then other animals such as horses, goats, cattle and water buffalo have been cloned.

Figure 6.21 Dolly the cloned sheep

Natural selection

Just as humans select individuals for further breeding, it has been found that selection in a habitat takes place naturally. Those individuals of a species which have the most suitable features to survive will continue to live and breed and pass on their features. Individuals which have features that do not equip them for survival will perish, and so do not pass on their features to future generations. This form of selection is used to explain the theory of evolution, where one species changes in time until another species is produced. The finches on the Galapagos Islands (Figure 6.22), first studied by Charles Darwin (1809–1882), are thought to have evolved by **natural selection**.

For discussion

How do you think that the human species may evolve in the future?

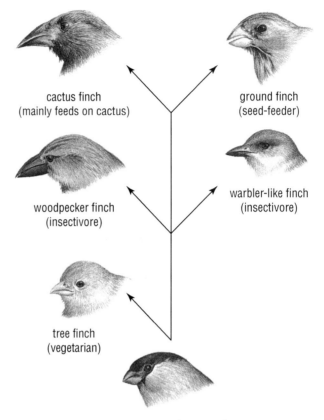

cactus finch
(mainly feeds on cactus)

ground finch
(seed-feeder)

woodpecker finch
(insectivore)

warbler-like finch
(insectivore)

tree finch
(vegetarian)

common ancestor insect-eating
and seed-eating ground finch

Figure 6.22 Darwin's finches

Charles Darwin

Charles Darwin (1809–1882) was born in England and as he grew up
he considered being either a doctor or a priest but after reading about
Alexander von Humboldt (1769–1859), a German naturalist, he decided
to become a naturalist too. A naturalist is a person who studies plants,
animals, rocks, landscapes and the weather.

For discussion

**Darwin was inspired to be a naturalist after reading about
Humboldt's activities which included traveling in Western Europe,
South America, North America and Russia, studying plants and
rocks, volcanoes, landscape and the Earth's magnetism making a
huge number of discoveries including the connection between the
Orinoco and Amazon rivers in South America.**

**Have you been inspired by others to take up an activity? Has the
work of a scientist or an investigation you have made inspired you
to take a further interest if only for a short time?**

Figure G Charles Darwin

Ideas on evolution before Darwin

When Darwin began his studies most scientists believed that the Earth had existed for only a few thousand years but the work of scientists called geologists, who studied rocks, was providing evidence that the Earth was much older. The Scottish geologist Charles Lyell (1797–1875) gathered the data of earlier geologists and made some observations of his own and wrote about how rocks and landscapes seemed to change slowly over long periods of time in a book called *Principles of Geology*. When Darwin read it, he too believed that the Earth was much older than most people thought.

Most scientists at the time also had a belief about living things. They believed in the ideas of the Ancient Greek Aristotle (384–322 BCE). Aristotle had looked at the living things around him and seen that some had simpler bodies than others. This led him to believe that living things could be grouped and placed on a ladder, called the ladder of nature, with the simplest on the bottom rung and the most complex – humans – on the top rung. Aristotle also believed that all the species had existed since the beginning of life on Earth and had remained unchanged throughout the Earth's history.

One scientist who doubted Aristotle's ideas was Comte de Buffon (1707–1788), a French naturalist. From his studies on plants and animals from round the world he believed that in the past plants and animals had developed in one place and then spread out and responded to the different environments by making changes to their bodies. He could not explain how the changes might come about.

Buffon's beliefs began to be taken up by other scientists. Erasmus Darwin (1731–1802), an English doctor, the grandfather of Charles wrote a book called *Zoonomia* in which he described how he thought life started with one species and then developed into others. He did not have any evidence to support his ideas.

Jean-Baptiste Lamark (1744–1829), a French naturalist also believed that one species developed into another and then another as conditions changed. He did not believe that a species could become extinct. He believed that the way a species changed was by developing a feature and then passing it on to future generations. At the time the giraffe had been newly discovered and he used it to explain his ideas. He thought that it had developed from a smaller antelope that stretched its neck, legs and tongue to feed in the tree tops. These stretched features were then passed onto the offspring and when they stretched to reach higher their necks, legs and tongue became even longer. This theory was known as the theory of acquired characteristics.

For discussion

Imagine that you lived in Greece by the sea like Aristotle. Look at pictures of sealife in rock pools and creatures brought in on fishing boats. Look at some pictures of land animals or land animals in your area. Arrange them into a ladder of nature. How reasonable does Aristotle's idea seem to you?

Figure H This picture of Comte de Buffon shows him with some of the living things he studied.

For discussion

Giraffes have blotches on their coat of fur to help them hide from predators. Can Lamark's theory of acquired characteristics explain how the giraffe got them? Explain your answer.

Darwin and his work

In 1831 Darwin was employed as a naturalist on a ship called the *Beagle*, which was about to make a journey around the world. The trip was originally planned to take two years but took five. The *Beagle* left England and sailed to South America then into the Pacific and on to the Galapagos Islands before sailing to Australia then across the Indian Ocean to the tip of South Africa then back to South America before returning home.

Darwin was mainly concerned with geological specimens on the trip but when he was on the coast of Chile he observed mockingbirds. Later, he visited the Galapagos Islands

Figure I Mockingbirds on a beach of one of the Galapagos Islands.

just over a thousand kilometres west of the Chilean coast and discovered more mockingbirds. He found that they all differed from those in Chile and also the mockingbirds on each island differed from each other. These observations made him seek an explanation and he thought that the species on the coast reached the Galapagos Islands but as it settled on each island it developed features to help it survive. He believed it provided evidence of evolution and knew from Lyell's work that the Earth could be very old so that there was time for this evolution to take place.

It was the custom on scientific expeditions in Darwin's time to collect specimens to bring home. This frequently meant shooting animals and preserving their bodies for the journey back. When Darwin returned to England he examined the specimens of finches collected on the Galapagos Islands and discovered that they too showed variations between islands, particularly in the size and shape of the beak and they resembled a species of finch found in Chile. He believed that they could have developed in the same way as the mockingbirds, but with these specimens the beaks gave him an idea about how evolution might take place. He reasoned that on each island the birds adapted to feeding on particular food. Adaptation and evolution became linked in his mind.

While Darwin was pondering his ideas of evolution he read a book called *An Essay on the Principle of Population* by Thomas Malthus (1766–1834) an Englishman who studied human societies. In the book Darwin learnt that a human population always increases faster than the food supply despite the best efforts of farmers and that the size of the population is eventually reduced by starvation and disease. Darwin thought that these population changes could occur in animals like his finches too and set out his thoughts in a similar way to the following points (page 109).

1 Animals and plants produce a large number of offspring. This number is far higher than the number of parents. For example a fly lays a large number of eggs and a plant such as the dandelion releases many wind-dispersed fruits.

2 The size of a plant or animal population in a habitat usually stays the same but a change in the habitat such as the removal of plants and exposing the soil may cause a change in the populations of plants as they grow back onto it.

3 If a living thing produces a large number of offspring yet only a few survive there must be a struggle for survival and this could be due to a limited supply of food or some other feature of the habitat.

4 When you look at a population of any living organism you can see that the individuals vary from each other.

5 In any species there is competition between the variety of individuals in the population. Those individuals which have the best features to survive in the conditions will have a better chance of surviving than those that have less suitable features. As a consequence the best-suited individuals will leave more offspring than the others and these may also have the favourable features. If this continues for some time a new species may develop and this process, which Darwin called natural selection, can explain how the different species of finch and mockingbird developed or evolved on the Galapagos Islands.

10 Was Darwin the first scientist to suggest that living things evolved? Explain your answer.

11 How is natural selection connected to evolution?

12 What evidence from a secondary source did Darwin use when thinking about the long time over which natural selection takes place?

13 What creative thought did Darwin have about the mockingbirds and finches on the Galapagos?

14 What evidence from a secondary source did Darwin use when thinking about the way in which the populations of finches changed?

15 What two pieces of evidence did Darwin consider to decide that there must be a struggle for survival among the individuals in a species?

16 What piece of evidence did Darwin consider to decide that new species are produced by natural selection?

◆ SUMMARY ◆

◆ Living things can be classified into groups (*see page 86*).

◆ Keys use features of a species that do not vary to help in identification (*see page 89*).

◆ Living things have characteristics which they inherit from the generation before them (*see page 94*).

◆ New varieties of species can be produced by selective breeding (*see page 101*).

◆ Natural selection takes place on the organisms in a habitat. It may cause one species to change into or evolve into a new species (*see page 105*).

End of chapter question

The sunflower is really a head of many smaller flowers, which fill up its centre and form dark coloured fruits that are usually called seeds. The seeds form a new generation of sunflower plants.

Plan an investigation to find out how the seeds vary from one another.

Figure 6.23

CHEMISTRY

The structure of the atom

◆ The Ancient Greeks' original idea of atoms
◆ Investigations of chemical reactions that supported the idea of atoms
◆ Dalton's atomic theory
◆ The structure of the atom first proposed by Joseph J. Thomson
◆ Ernest Rutherford's investigations revealing the basic structure of the atom
◆ Atomic structure
◆ The atomic structure of the first 20 elements in the periodic table

We have seen that biologists investigated the structure of plants and animals and found them to be composed of cells. Later, when it was possible to build more powerful microscopes, the structure of the cells was discovered too. Chemists have also investigated the structure of matter and discovered that it is composed of particles called **atoms**. Later, as new apparatus was developed, it became possible to go further and investigate the structure of atoms.

Discovering the structure of the atom

For the greatest part of human history, people were too busy finding food and shelter and staying alive to think about how they and their surroundings were made. With the development of farming and the regular production of food, it became possible for towns to develop and for some people to earn a living teaching others about what was known about the world. The teachers of Ancient Greece were called philosophers, and from their observations they generated ideas about almost everything, which they taught to anyone who was willing to listen.

The painting shown in Figure A is thought to depict every one of the major Ancient Greek philosophers, although in fact many of them lived at different times. It shows how the philosophers argued to make others pay attention to their ideas.

Some of the philosophers used myths and legends to explain their surroundings but Thales (about 624–546 BCE) was different. He looked around him and used what he saw to try to explain the world. Thales lived in Miletus on the coast of the Mediterranean so water was nearly always in view. This made Thales think that perhaps the world had formed from water. Anaximenes (about 585–528 BCE), who was a student of Thales, disagreed with him. He thought that as everything was surrounded by air then everything was made from air.

Figure A *The School of Athens* by Raphael is a fresco at the Vatican in Rome. Plato (428–348 BCE) and his pupil Aristotle (384–322 BCE) are shown in the centre.

For discussion

If you were a philosopher living
a) **in a forest**
b) **in the Arctic**
c) **on the side of a volcano**
what would you tell your students the world was really made of? How would you explain the world to them in the terms of your answers?

The idea of atoms

Democritus (about 460–370 BCE) decided to argue against these ideas and thought about what would happen if you cut something into smaller and smaller pieces until you could not cut it any smaller. He concluded that you would come to an indivisible particle, which he called an 'atom'. Building on this idea that everything was not made of water or air but of atoms, he went on to say that atoms of different substances had different sizes and shapes, and that they were able to join together to make even more different substances.

The other philosophers could not agree on Democritus' creative thought and later Aristotle put forward the idea that matter was made from water, air, fire and earth! This idea was believed to be true until scientists began investigating chemical reactions.

1 Why do you think Thales has been called the 'Father of Science'?

2 Do you think Democritus should be called the 'Father of Science' instead? Explain your answer.

Investigations on chemical reactions

Antoine Lavoisier (1743–1794), a French chemist, investigated the changes that took place when two chemicals reacted and formed a new compound. He weighed the chemicals before the reaction and then weighed the compound that was formed. Lavoisier found that the total mass of the chemicals was the same as the mass of the compound that was produced. From this result and from the results of similar experiments, Lavoisier set out his law of conservation of mass, which stated that matter is neither created nor destroyed during a chemical reaction.

Lavoisier was assisted in his work by his wife Marie-Anne who made sketches of the new pieces of apparatus that were devised for the investigations and translated the work of other scientists into French for Antoine to read.

Joseph Proust (1754–1826), another French chemist, followed Lavoisier's example by carefully weighing the chemicals in his experiments. He discovered that when he broke up copper carbonate into its elements — copper, carbon and oxygen — and then weighed them, they always combined in the same proportions of five parts of copper, four parts of oxygen and one part of carbon. He found that other substances were made from different proportions of elements and these proportions were always the same too, no matter how large or how small the amounts of elements that were used. From his work, Proust devised the law of definite proportions, which stated that the elements in a compound are always present in a certain definite proportion, no matter how the compound is made.

Dalton's atomic theory

John Dalton (1766–1844) was an English chemist who studied gases and from his investigations on the combining of carbon and oxygen he produced two gases. The first of them seemed to be made from one particle of carbon joining with one particle of oxygen, and in the second gas it seemed that one particle of carbon joined with two particles of oxygen. From his own observations such as these, and from reading about the work of Lavoisier and Proust, Dalton put together his atomic theory. He suggested that:

1 all matter is composed of tiny particles called atoms
2 atoms cannot be divided up into smaller particles and cannot be destroyed
3 atoms of an element all have the same mass and properties
4 the atoms of different elements have different masses and different properties
5 atoms combine in simple whole numbers when they form compounds.

This theory helped chemists at the time, but the results of later investigations showed that it was not completely correct, as we shall see.

The plum pudding atom

The scientists in these early chemical studies on atoms used a unit called the 'atomic weight' to

Figure B This 1788 portrait shows the French chemist Antoine Lavoisier and his wife Marie-Anne Pierrette Paulze (1758–1836), who helped her husband with much of his scientific work.

3 The gas made from molecules each containing one carbon atom and one oxygen atom is called carbon monoxide. What is the name of the second gas Dalton investigated?

4 Dalton used evidence from the work of others to help him develop his theory.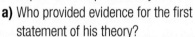
   a) Who provided evidence for the first statement of his theory?
   b) Who provided evidence for the second statement of his theory?
   c) What two pieces of evidence helped him develop the fifth statement of his theory?

compare the elements. Today we use the term relative atomic mass or RAM. An English chemist called William Prout (1785–1850) studied the atomic weights of the different elements and thought he could use them to explain the structure of atoms. He knew that hydrogen had the lowest atomic weight and the atomic weights of all the other elements appeared to be multiples of the atomic weight of hydrogen. This suggested to him that all the other elements were made up from different numbers of hydrogen atoms. This idea was later shown to be completely wrong but it did make scientists such as Joseph J. Thomson think that atoms might have a structure inside them.

During the 19th century, great developments were made in the study of electricity and the development of electrical apparatus for use in investigations. One of these pieces of apparatus was the cathode ray tube, which produces rays when it is connected into an electrical circuit. The ray is produced from the material from which the cathode is made. Thomson investigated these rays and discovered that they were made of tiny particles, which had a mass over a thousand times smaller than a hydrogen atom. Thomson called the particles 'corpuscles' but George Stoney (1826–1911), an Irish physicist, named them electrons. When he used different materials for the cathode he always found that the electrons they produced were the same.

Figure C Joseph J. Thomson at work in his laboratory

In 1904 Thomson devised a model of the structure of the atom from his studies on electrons. He proposed that electrons were present in the atom. He knew that electrons were negatively charged and atoms were neutral, so the negatively charged electrons must be balanced by a positive substance in the atom. He described the atom as being like a plum pudding with the negatively charged electrons being surrounded by a positively charged 'pudding'.

Ernest Rutherford and the atom

Ernest Rutherford (1871–1937) was born and raised in New Zealand and, after his successful studies on electricity and magnetism at the University of New Zealand in Wellington, he moved to Cambridge University in England to work with J. J. Thomson. He spent time studying radioactive materials and the radiation that they produced with Paul Viliard (1860–1934), a French scientist. Between them they discovered that there are three types of radiation – alpha particles, beta particles and gamma rays.

Rutherford had found that alpha particles were large, positively charged particles much larger than electrons and decided to use them to test Thomson's idea about the plum pudding structure of the atom.

Rutherford's plan was to hang up a thin sheet of gold and surround it with a screen that could detect alpha particles, as shown in Figure D (page 116). He would then fire alpha particles at the sheet and the alpha particles would eventually hit the screen and be detected. From the marks made by the alpha particles on the screen he could work out the structure of the atom. Rutherford predicted that if Thomson's model was correct all the alpha particles would pass straight through the gold atoms and make a mark directly behind the gold sheet.

The experiment was set up and the alpha particles were fired at the gold sheet. Most of the alpha particles made a mark directly behind the metal sheet but some made marks all round the screen. This did not fit in with the prediction and suggested that the atoms had a structure that was not like a plum pudding.

5 Why could Prout's idea be considered a creative thought?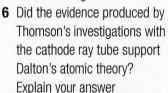

6 Did the evidence produced by Thomson's investigations with the cathode ray tube support Dalton's atomic theory? Explain your answer

7 What creative thought did Thomson have after his discovery of electrons?

8 What do you think the plum pudding model looked like? Draw and label it.

Rutherford reasoned that as some alpha particles appeared all over the screen they must be hitting and 'bouncing off' something inside the atoms that repelled them, but, as most passed through, there must be a large amount of empty space in an atom to let the alpha particles through.

After further thought, Rutherford concluded that an atom did not have a positively charged 'pudding' around the outside, but instead had a positively charged centre or nucleus, which was surrounded by negatively charged electrons.

Rutherford's later work identified particles in the nucleus which he named protons after Joseph Proust who first suggested that atoms might have smaller particles inside them.

When Rutherford looked at all the evidence that had been collected about atomic structure, he found that there seemed to be something missing to explain all the results. He thought that there must be other particles present in the nucleus which are similar to protons but do not have an electrical charge. He thought that, for a structure to have no overall charge, it must be made of a positively charged proton and a negatively charged electron together, and called the particle a 'neutral doublet'.

This critical look at the data led other scientists to evaluate the methods used for investigating atomic structure and refine them for further investigations. In 1932, an English scientist named James Chadwick (1891–1974) fired alpha particles at beryllium atoms and knocked out particles that had a similar mass to protons but no electrical charge. He had discovered the neutral doublet predicted by Rutherford, and he called it the neutron. Further work on this particle by others showed that it was not composed of a proton and an electron –rather, it was a particle with a similar structure to a proton but without the electrical charge.

9 What scientific enquiry skill was Rutherford using in selecting Thomson's model for investigation?

10 What apparatus did Rutherford select for his investigation?

11 What observations were necessary in the investigation?

12 What did Rutherford predict if Thomson's model was correct?

13 How did the results compare with the prediction?

14 What creative thought did Rutherford have to explain the results?

15 What scientific knowledge and understanding did Rutherford use to describe the structure of the neutral doublet?

16 Assess the accuracy of Rutherford's prediction about a neutral particle.

The experiment

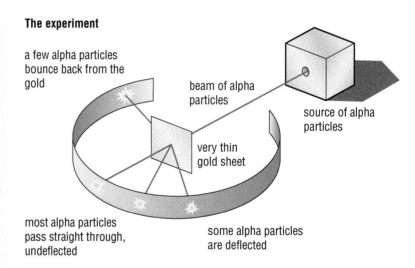

a few alpha particles bounce back from the gold

beam of alpha particles

source of alpha particles

very thin gold sheet

most alpha particles pass straight through, undeflected

some alpha particles are deflected

Explanation

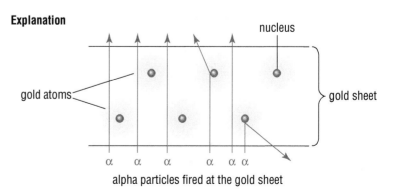

nucleus

gold atoms

gold sheet

α   α   α   α   α α

alpha particles fired at the gold sheet

Figure D Rutherford's experiment investigating atomic structure

The structure of atoms

An atom is about a ten-millionth of a millimetre across. It is made of **subatomic particles** (Figure 7.1). At the centre of the atom is the nucleus. This is made from two kinds of subatomic particles called **protons** and **neutrons**. (Hydrogen is an exception because it has only a proton in its nucleus.) A proton has the same mass as a neutron. It also has a positive electrical charge, while a neutron does not have an electrical charge.

Around the nucleus are subatomic particles called **electrons**. Each electron has a negative electrical charge and travels at about the speed of light as it moves around the nucleus.

The number of electrons around the nucleus of an atom is the same as the number of protons in the nucleus. The negative electrical charges on the electrons are balanced by the positive electrical charges on the protons. This balancing of the charges makes the atom electrically neutral – it has no overall electrical charge.

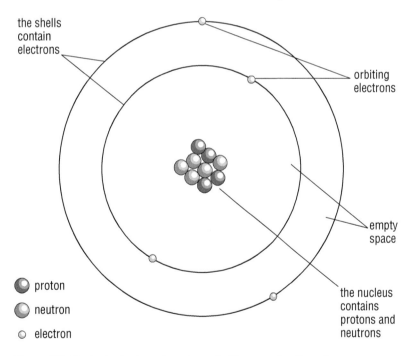

Figure 7.1 The basic structure of an atom – in this case, a beryllium atom

The electrons are arranged in groups at different distances from the nucleus. They are described as being arranged in shells. For example, the carbon atom has two electrons close to the nucleus making an inner shell and four electrons further away making an outer shell.

Many atoms have more shells than this. For example, the lead atom has six shells (Figure 7.2).

All the atoms in each element have the same number of protons. For example, carbon atoms always have six protons and sodium atoms always have eleven protons.

The number of neutrons in the atoms of an element may vary. Most carbon atoms, for example, have six neutrons but about 1% of carbon atoms have seven neutrons and an even smaller proportion of carbon atoms have eight neutrons. These atoms of an element that have different numbers of neutrons are called isotopes.

1 How many electrons are there in each shell of an atom of lead?

2 What are isotopes?

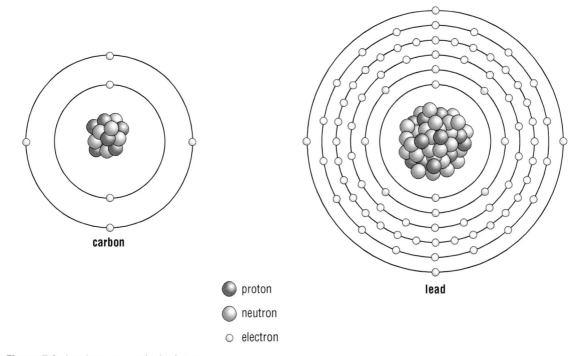

carbon

⬤ proton

◯ neutron

○ electron

lead

Figure 7.2 A carbon atom and a lead atom

The atomic structure of 20 elements

For discussion

How many of the atoms in Figure 7.3 can you recognise by their symbols? You can check your answer by looking at part of the periodic table in Figure 8.2 on page 125.

In *Student's Book 2* you learnt the names and symbols of the 20 lightest elements in the periodic table. Figure 7.3 shows the structures of those 20 elements for you to compare. Notice that the detail of the atomic nucleus is shown as the number of protons (p) and neutrons (n) at the centre of each diagram of the atom and that an electron in the shell of an atom is shown as (e).

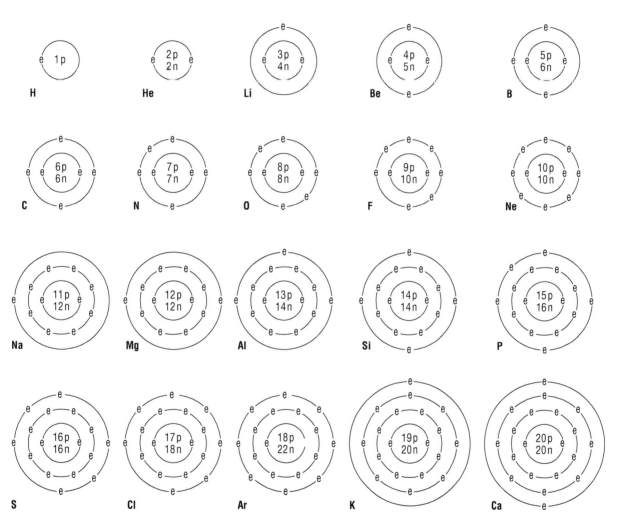

Figure 7.3 The atomic structure of the first 20 elements of the periodic table

◆ SUMMARY ◆

◆ The idea of atoms came from the work of Ancient Greek philosophers (*see page 113*).

◆ Investigations on chemical reactions supported the idea of atoms (*see page 113*).

◆ Dalton's atomic theory was derived from his own work and the work of others (*see page 114*).

◆ William Prout thought that atoms were made up from hydrogen atoms and his idea led to investigations on atomic structure (*see page 114*).

◆ J. J. Thomson proposed the first structure of the atom, known as the plum pudding model (*see page 114*).

◆ Ernest Rutherford's investigations revealed that the atom has a nucleus surrounded by a cloud of electrons (*see page 115*).

◆ James Chadwick discovered the neutron (*see page 116*).

◆ An atom is composed of a central nucleus of protons and neutrons (hydrogen atoms only have one proton and no neutrons) and a cloud of electrons (*see page 117*).

◆ The atoms of the first 20 elements increase by one proton and one electron for each position across the rows of the periodic table (*see page 119*).

End of chapter questions

Look at Figure 7.3 on page 119 and answer the following questions.

1 Look at the number of protons in the atoms. Can you see a pattern in their number as the atoms increase in size? Explain your answer.

2 Look at the number of neutrons in the atoms. Can you see a trend as you move from the smallest atom to the largest?

3 Compare how the numbers of protons and neutrons change as you move from the smallest to the largest atom.

8 The periodic table

◆ How 19th century scientists used data about elements to sort them into order
◆ How modern scientists group the elements in the periodic table
◆ The alkali metals – Group 1 of the periodic table
◆ The alkaline earth metals – Group 2 of the periodic table
◆ The halogens – Group 7 of the periodic table
◆ The noble gases – Group 8 of the periodic table
◆ The unique properties of hydrogen

When large amounts of scientific data are produced there is a need to sort the data out so that different parts can be easily found when needed for further study. In the last chapter we saw how both biologists and chemists performed investigations to find the structure of cells and atoms respectively. In Chapter 6 we looked at how biologists sorted out the diversity of living things by looking at their features and putting them into groups. In this chapter we will look at how chemists sorted out the elements into groups and the characteristics of the elements in those groups.

Figure 8.1 Baron Justus Freiherr von Liebig working in his laboratory in the 19th centrury.

For discussion

How does Liebig's laboratory differ from your school laboratory? Would you expect great discoveries to be made in a laboratory like this one? Explain your answer.

Scientists in the 19th century worked in laboratories such as the one shown in Figure 8.1, and made many discoveries about elements and compounds. This picture shows Baron Justus Freiherr von Liebig (1803–1873), a German chemist who did not contribute to sorting out the elements but discovered the structure of many compounds and designed apparatus such as the Liebig condenser, which you may have used in distillation.

Sorting out the elements

John Dalton, the English chemist who constructed the atomic theory (page 114), also tried to put the elements in order. He measured the masses of the elements he collected when he broke up compounds. At the time, the term 'mass' was not used. The terms 'weight' and 'atomic weight' were used instead and they are used in this feature, and in the feature on Mendeleev, as they help to show how chemists thought in the 19th century.

Over time there were many revisions of the idea of atomic weights and today chemists use the term relative atomic mass in their chemical calculations.

Dalton used the weight of hydrogen to compare with the weight of other elements. For example, when he separated hydrogen and oxygen from the compound water, he found that the weight of oxygen was seven times greater than the weight of hydrogen. As he believed that one atom of hydrogen combined with one atom of oxygen to make a molecule of water, he thought that the atomic weight of hydrogen was one and the atomic weight of oxygen was seven. He used this idea to measure the atomic weights of other elements and set them out in a table.

Unfortunately Dalton was not a very accurate experimenter and other scientists found that the weight of oxygen produced when water splits up is eight times greater than the weight of hydrogen, so they thought that its atomic weight should be eight. However, Dalton had also made a mistake in thinking that all atoms combine in the ratio of 1 : 1. It was later discovered that a molecule of water contains two atoms of hydrogen and one atom of oxygen. This means that the weight of an oxygen atom is eight times the weight of two atoms of hydrogen, or sixteen times the weight of one hydrogen atom – making its atomic weight 16, not 8.

Johann Wolfgang Dobereiner (1780–1849), a German chemist, studied Dalton's work. In 1829, when more than 12 new elements had been discovered, he began sorting them out and found that he could divide them into groups of three according to their atomic weights and properties. These groups became known as Dobereiner's Triads. In one triad he placed lithium, sodium and potassium and in another he placed chlorine, bromine and iodine, while in a third triad he placed calcium, strontium and barium.

Figure A The symbols and atomic weights of substances that Dalton believed to be elements in 1805

1 'Azote' is another name for nitrogen and 'platina' is another name for platinum but there are six names in Dalton's list that are not elements but are compounds. Which are they?

Another ten elements had been discovered by the time the English chemist John Newlands (1834–1907) attempted to sort out the elements. Newlands set out the elements in order of atomic weights, starting with the lowest. As he examined the properties of the elements in his list he noticed that some of the properties appeared periodically in elements that were eight places apart. From this he devised a 'law of octaves' to explain his discovery and compared it to the arrangement of notes in an octave of music. This was unfortunate because other scientists thought comparing science and music in this way was ridiculous but in time they accepted that his discovery had been of some value in sorting out the elements.

2 What was the inaccurate observation that Dalton made?

3 What inaccurate conclusion did Dalton make from his studies?

4 In what way was Dalton's work useful as evidence to Dobereiner and Newlands?

5 Could Dobereiner and Newlands attempt a more detailed sorting out of the elements than Dalton because they had more data? Explain your answer.

6 Set out the elements along a line in order of atomic weight, using Figure 7.3 on page 119 to help you. Start the line with lithium (Li) and finish with calcium (Ca), but miss out neon (Ne) and argon (Ar), which were not known when Newlands was working. Mark the positions of lithium, sodium and potassium, which have similar properties. Count the number of elements from lithium to sodium, and also from sodium to potassium. What do you find?

7 How do you think the discovery of neon and argon affected the law of octaves?

Figure B John Newlands noticed patterns in the properties of the elements, when arranged in order of atomic weight.

Dimitri Mendeleev

Dimitri Mendeleev was born in 1834 in Tobolsk, Siberia, which is in Russia. He was the youngest of a large family of over ten children. His father was headteacher of a local school but died while Mendeleev was young. His mother's family owned a glass factory and after his father's death she took over the running of it. This allowed the young Mendeleev to visit the factory and learn about the science of glass making from the factory chemist and the art of glass making from the glass blower. When Mendeleev was 14, the glass factory was destroyed in a fire and there was no money to rebuild it. Mendeleev's mother decided that he should go to university. In order to do this, he needed to do well in subjects such as history and Latin but he was only interested in science. His mother and a friend who was a teacher encouraged him to work harder in these subjects and eventually he reached university and studied chemistry.

His studies went well until the third year, when he became so ill that he could not leave his bed. However, he insisted on keeping up the work and his professors set him assignments which his fellow students brought

Figure C Mendeleev as a young man working in a laboratory →

to him. He completed his work on time and received a medal for being the top of his class. In time, Mendeleev recovered and travelled in France and Germany to find out more about the latest developments in chemistry. He worked with many distinguished scientists including the German chemist Robert Bunsen, after whom the gas burner was named because he used it a great deal. After Mendeleev returned to Russia, he became the Professor of Chemistry at the university in St Petersburg and dedicated his life to the study and teaching of chemistry.

Following on from the work of John Newlands, Mendeleev began to sort out the elements. He arranged them in order of their atomic weights but he also noticed something else – about the way the atoms joined together. After Dalton's error in thinking that atoms only join in a ratio of 1 : 1, it had been discovered that the atoms of some elements were capable of joining with two, three or more atoms of other elements. When Mendeleev studied the elements in his list, he noticed that lithium would join with only one other atom but beryllium, next to it, would join with two atoms. The next element, boron, joined with three atoms and carbon, which followed, joined with four atoms. As he went down his list he saw that the elements joined with three, two and one atoms respectively before there was another rise and fall and so on.

Figure D Handwritten early version of Mendeleev's periodic table – you can see how he made changes as new data became available.

Mendeleev rearranged the elements in his list into rows in a table so that each row had a rise and fall in the number of atoms the elements would combine with. When he looked at the properties of the elements as they were arranged one below the other in columns in the table, he noticed that they not only combined with the same number of atoms but also had similar properties. As he looked along the rows in his table he could see that the ability of the elements to join with atoms and the properties they possessed changed periodically and this led to the table being named the periodic table.

However, there was a problem with the table – all the data did not fit in to create a regular pattern in some places. Mendeleev knew that many elements had been discovered in the 19th century and there was a chance that there were more still to be discovered. He therefore left gaps in his table that could be filled as the elements were discovered. This meant that he could use the table in another way – to look at the elements around the gaps and predict the properties of the unknown elements. Other scientists did not approve of the idea of leaving gaps. They thought that it was cheating and that it was done to support his theory, which they found unbelievable. However, they had to change their minds when new elements were discovered that had the properties Mendeleev predicted and the elements were added to the table.

Mendeleev is best known for the periodic table but in Russia he was well known as a very enthusiastic teacher and lecturer and would even gather people around him while he travelled on trains and tell them about chemistry. He wrote books on chemistry and made investigations in agriculture, the extracting of minerals from rocks and the refining of oil to maximise its usefulness. He investigated the properties of gases and used a hot air balloon to rise into the sky to observe an eclipse of the Sun.

His fame for developing the periodic table spread around the world and he received many awards for it during the rest of his life. He died in 1907 aged 73.

8 Curiosity is a sign of a scientist. How did Mendeleev show this in his early life?

9 How did Mendeleev show his curiosity in later life?

10 When did Mendeleev show great perseverance?

11 Was Mendeleev a good communicator of science? Explain your answer.

12 What evidence did Mendeleev consider when starting work on the periodic table?

13 How did Mendeleev's powers of observation help him in constructing the periodic table?

14 What creative thought did Mendeleev apply to the table when he found that some of the data did not fit to form a regular pattern?

15 How did Mendeleev test his table to see if it could be used to show the relationships of all elements?

The periodic table

Earlier we saw that chemists in the 19th century used the term 'atomic weight' in sorting out the elements (page 122) and that atomic weights were used by Mendeleev to help him construct the periodic table. Over the years the periodic table has been revised and today the elements are arranged in order of atomic number.

In Figure 7.3 (page 119), we saw that the atoms of each element have a nucleus with a certain number of protons, unique to that element. The number of protons in an atom is called the **atomic number** and this is now used to arrange the elements in the periodic table. The atomic number of each element can be seen in the top left corner of the box for that element.

1 How are the elements arranged by their atomic number – horizontally or vertically?

Figure 8.2 Part of the modern periodic table

Groups of the periodic table

Many of the columns of elements in the periodic table are called **groups**. The elements in a group share similar properties. A trend can be seen in the properties as you go down the group.

Group 1, the alkali metals

In this group, the metals themselves are not alkalis, but the oxides and hydroxides that they form are. It is this property of these compounds that gives the metals in this group their name.

Physical properties of the alkali metals

Table 8.1 shows some of the physical properties of the alkali metals.

Table 8.1 Physical properties of the alkali metals

Element	Density/g/cm³	Melting point/°C	Boiling point/°C
lithium	0.53	180.6	1344
sodium	0.97	97.9	884
potassium	0.86	63.5	760

Chemical properties of the alkali metals

When samples of the alkali metals are added to water in turn, the following reactions take place. Lithium fizzes as it floats on the water, sodium fizzes more strongly and potassium bursts into flame. In all three reactions, the metal hydroxide is produced in the liquid and hydrogen gas escapes into the air. In the reaction with potassium and water, the heat produced causes the hydrogen to burn in the air and produce a flame (Figure 10.4, page 148).

When the alkali metals burn in air they produce solid metal oxides (Figure 10.2, page 146). They react vigorously with elements in group 7, the halogens, to produce halides.

Uses of the alkali metals

- **Lithium** – Lithium's name is derived from *lithis*, the Greek word for stone, because it is found in many kinds of igneous rock. It is used in batteries and in compounds used as medicines to treat mental disorders.

- **Sodium** – Metallic sodium is used in certain kinds of street lamps that give an orange glow. It is alloyed with potassium to make a material for transferring heat in a nuclear reactor. Sodium compounds such as sodium hydroxide have a wide range of uses. In the body, sodium ions are needed by nerve cells to transfer electrical signals called nerve impulses.
- **Potassium** – Potassium compounds such as potassium nitrate are mined and used as fertiliser. In the body, potassium ions are used for the control of the water content of the blood and also used, with sodium ions, in sending electrical signals along nerve cells.

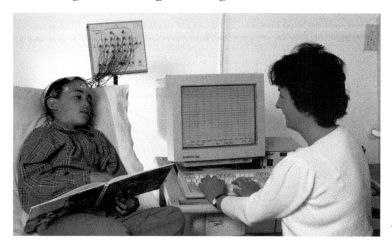

Figure 8.3 Measuring brain waves – nerve impulses are due to the movement of sodium and potassium ions in nerve cells.

2 Which of these statements about the trends shown in Table 8.1 are true?
  a) Down the group, the density of the metals generally:
  - increases
  - decreases.
  b) Down the group, the melting point of the metals generally:
  - increases
  - decreases.
  c) Down the group, the boiling point of the metals generally:
  - increases
  - decreases.
3 Sodium is a softer metal than lithium. Describe how you think the softness of potassium compares with that of sodium.
4 Which metal in Table 8.1 has the smallest temperature range for its liquid form?
5 How does the strength of the reaction with water vary as you move down the group of alkali metals?
6 From the way alkali metals behave with water, how do you think the strengths of their reactions with oxygen and the halogens vary as you move down the group?

Group 2, the alkaline earth metals

These metals are not alkalis but their oxides and hydroxides dissolve slightly in water to make alkaline solutions.

Physical properties of the alkaline earth metals

Table 8.2 shows some of the physical properties of these metals.

Table 8.2 Physical properties of the alkaline earth metals

Element	Density/g/cm³	Melting point/°C	Boiling point/°C
beryllium	1.85	1289	2476
magnesium	1.74	649	1097
calcium	1.53	840	1493

For discussion

How do the trends shown in Table 8.2 compare with those shown in Table 8.1?

Chemical properties of the alkaline earth metals

The alkaline earth metals behave with oxygen in the air in the following ways. Beryllium does not react with oxygen in the air, while magnesium, calcium and strontium form metal oxides on their surfaces, and barium must be stored in oil to prevent a vigorous reaction. All the elements burn in pure oxygen to produce oxides. Beryllium does not react with water or even steam. The other alkaline earth metals do react with water but their reactions are not as strong as those between water and the alkali metals.

Uses of the alkaline earth metals

- **Beryllium** – Beryllium combines with aluminium, silicon and oxygen to make a mineral called beryl. Emerald and aquamarine are two varieties of beryl that are used as gemstones in jewellery. Beryllium is mixed with other metals to make alloys that are strong, yet light in weight. It is also used in a mechanism that controls the speed of neutron particles in a nuclear reactor.
- **Magnesium** – Magnesium is used in fireworks to make a brilliant white light. Another important use is to mix it with other metals to make strong, lightweight alloys such as those used to make bicycle frames.

Green plants need magnesium in order to make the chlorophyll that traps the energy from sunlight in photosynthesis. Magnesium is needed in the body for the formation of healthy bones and teeth.

- **Calcium** – Calcium's name is derived from the word *calx*, which is the Latin name for lime (calcium oxide). Calcium forms many compounds with a wide range of uses, from baking powders and bleaching powders to medicines and plastics. In the human body, calcium is required for the formation of healthy teeth and bones and for the contraction of muscles.

Figure 8.4 The green mineral is emerald, which contains the metal beryllium. Emerald can be cut and polished to make beautiful jewellery.

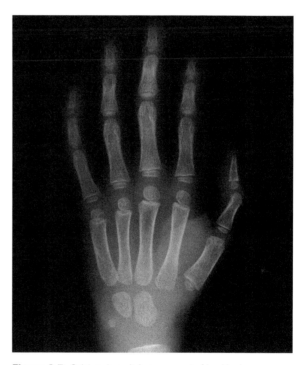

Figure 8.5 Calcium is a vital component of healthy bones.

7 Which of these statements about the trends shown in Table 8.2 are true?
   a) Down the group, the density of the metals generally:
   - increases
   - decreases.
   b) Down the group, the melting point of the metals generally:
   - increases
   - decreases.
   c) Down the group, the boiling point of the metals generally:
   - increases
   - decreases.

8 Is there a trend in the way alkaline earth metals take part in chemical reactions? Explain your answer.

Group 7, the halogens

The word 'halogen' is a Greek word for salt former, and all the elements in this group form salts readily.

Physical properties of the halogens

Table 8.3 shows some of the physical properties of these elements.

Table 8.3 Physical properties of the halogens

Element	Melting point/°C	Boiling point/°C
fluorine	−219.7	−188.2
chlorine	−100.9	−34.0
bromine	−7.3	59.1

Chemical properties of the halogens

Fluorine is one of the most reactive elements and even forms compounds with the noble gases (group 8), which are very unreactive elements. When fluorine and water meet a vigorous reaction takes place in which oxygen and hydrogen fluoride are produced. Chlorine dissolves in water and reacts with it to produce hydrochloric acid and hypochlorous acid. Bromine also dissolves in water but reacts more slowly to form hydrogen bromide and hypobromous acid. Iodine does not dissolve in water and does not react with it.

Uses of the halogens

Figure 8.6 Fluorite is a mineral containing fluorine, which glows weakly when ultraviolet light is shone on it. This property is called fluorescence.

- **Fluorine** – Fluorine is a pale yellow-green poisonous gas. It is found in combination with calcium in the mineral fluorite (Figure 8.6). One variety of fluorite called Blue John has coloured bands and is carved to make ornaments and jewellery. Fluorine is combined with hydrogen to make hydrogen fluoride, which dissolves glass and is used in etching glass surfaces. Sodium fluoride prevents tooth decay and is added to some drinking water supplies. Fluorine is one of the elements in chlorofluorocarbons or CFCs.
- **Chlorine** – Chlorine is a yellow-green poisonous gas. It is found in combination with sodium as rock salt. Chlorine is used to kill bacteria in water supply systems and is also used in the manufacture of bleach. It forms hydrochloric acid, which has many uses in industry.

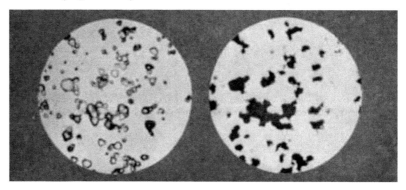

9 What trends can you see in the melting points and the boiling points of the halogens?

10 Over how many degrees Celsius is chlorine a liquid?

11 Which halogen is a liquid at a room temperature of 20 °C? Explain your answer.

12 From the information about halogens and water, what is the trend as you go down the group? Explain your answer.

13 How does the reactivity of the halogens and the alkali metals compare as you go down each group?

- **Bromine** – Bromine is a red-brown liquid, which produces a poisonous, strong-smelling brown vapour at room temperature. Bromine is extracted from bromide salts in seawater and is used, with silver, in traditional photography. Silver bromide is light sensitive and is used in photographic film to record the amount of light in different parts of the image focused by the camera lens (Figure 8.7).

Figure 8.7 Magnified images showing silver bromide crystals on a piece of photographic film (left) and silver deposits on a developed film (right).

Group 8, the noble gases

The noble gases are very unreactive. Some of their uses are listed below.

- **Helium** – Helium is lighter than air and is used to lift meteorological balloons into the atmosphere. These balloons carry equipment for collecting information for weather forecasting and relay it by radio to weather stations. Helium is also mixed with oxygen to help deep-sea divers breathe underwater.

Figure 8.8 Launching a meteorological balloon filled with helium gas.

- **Neon** – This gas produces a red light when an electric current flows through it and is used in lights for advertising displays.

Figure 8.9 Many of the advertising displays in Times Square, New York, use neon-filled light tubes.

- **Argon** – Argon is used in wire-filament light bulbs. When an electric current flows in the tungsten wire in the filament, the metal gets hot. If oxygen were present it would react with the hot tungsten and the filament would quickly become so thin that it would break. Argon is used instead of air containing oxygen because it does not react with the tungsten and the filament lasts longer. Argon is also used in making silicon and germanium crystals for the electronics industry.
- **Krypton** – This is used in lamps that produce light of a high intensity, such as those used for airport landing lights and in lighthouses.
- **Xenon** – Xenon is used to make the bright light in a photographer's flash gun.

Hydrogen

When you look at the periodic table (Figure 8.2) you can see that hydrogen is placed alone. The reason for this is that its properties do not match well with the properties of the other elements. Hydrogen can be considered to be unique.

The hydrogen atom is unusual in not having any neutrons. It has just one proton and one electron and this makes it the lightest atom. Hydrogen is a colourless gas without any smell. It is the most common element in the universe.

Hydrogen forms many compounds. It is found in acids and released when acids react with some metals. It is found in bases such as hydroxides and hydrogencarbonates. Hydrogen combines with carbon to make the hydrocarbons found in oil and is combined with nitrogen to make ammonia for use in fertilisers.

Hydrogen can burn when heated. If it is mixed with air or oxygen before it is ignited, it explodes and can cause a great deal of damage. The 'pop' that is used in the test to identify hydrogen gas is a very small explosion.

Despite the danger of explosions, cars have been developed that can run by burning hydrogen instead of petrol. The hydrogen that is needed as fuel is compressed into a tank and carefully released into the engine. The product of burning hydrogen is water vapour. There is no carbon dioxide produced, unlike when fossil fuels are burnt.

14 Construct a word equation for the burning of hydrogen.
15 Draw a diagram of a hydrogen atom.

For discussion

What precautions do you think should be taken if hydrogen-powered vehicles were to replace petrol-fuelled vehicles?

The periods of the periodic table

The horizontal rows of the periodic table are called **periods**. There is just hydrogen and helium in the first period, then eight elements in periods 2 and 3. In periods 4 to 6 there is a row of ten transition metals between the elements in groups 2 and 3. Many of these elements are coloured, for example copper and gold, or they produce coloured compounds – for example, the compounds of colbalt, nickel and iron (Figure 11.2, page 154). As you look along periods 2 to 6 you can see that the elements change from being metals to being non-metals.

16 Look at the elements in period 3 of the table in Figure 8.2 (from sodium to argon) and find the arrangement of their electrons in Figure 7.3 on page 119.
  a) How does the number of protons in the nuclei change as you move from left to right along the period?
  b) How does the number of electrons in the outer electron shell change as you move from left to right?

◆ SUMMARY ◆

◆ Dalton made a contribution to the sorting out of the elements (*see page 122*).

◆ The work of Dobereiner and Newlands built on the work of Dalton (*see pages 122–123*).

◆ Mendeleev produced the periodic table to show the relationships between the elements (*see page 123*).

◆ The periodic table shows how scientists arrange the elements into order today (*see page 125*).

◆ Columns in the periodic table are called groups (*see page 126*).

◆ Group 1 of the periodic table contains the alkali metals (*see page 126*).

◆ Group 2 of the periodic table contains the alkaline earth metals (*see page 128*).

◆ Group 7 of the periodic table contains the halogens (*see page 130*).

◆ Group 8 of the periodic table contains the noble gases, which are very unreactive (*see page 131*).

◆ Hydrogen has unique properties (*see page 133*).

End of chapter question

'*Elements in the groups of alkali metals, alkaline earth metals, noble gases and halogens are important in our lives.*'

Do you agree? Explain your answer.

Endothermic and exothermic reactions

◆ Endothermic reactions, including melting, the reaction of sherbet in the mouth, cooking, the breakdown of limestone and photosynthesis
◆ Exothermic reactions, including burning, respiration and oxidation
◆ Investigating a burning candle
◆ The danger of incomplete combustion
◆ Measuring energy in a fuel
◆ Improving fuel efficiency

An **endothermic reaction** is one in which heat is taken in. An **exothermic reaction** is one in which heat is given out. In this chapter we will look at examples of both kinds of reaction.

Endothermic reactions

Melting

1 How do the particles in ice change during the process of melting?

For discussion

Are evaporating and boiling endothermic processes? Explain your answer.

Probably the most familiar endothermic process is melting. Once the temperature of the surroundings rises above 0 °C, ice absorbs heat energy from the air and starts to melt. This is a reversible physical reaction, not a chemical reaction.

Figure 9.1 When ice melts, energy is taken in from the surroundings.

Sherbet

Sherbet is a popular sweet. It is made from citric acid and sodium hydrogencarbonate. When you put sherbet in your mouth, it feels cool. This is due to an endothermic reaction taking place which results in heat being taken from your body. The reaction occurs when the sherbet dissolves in your saliva and the two chemicals react together. The word equation for the reaction is:

$$\text{citric acid} + \text{sodium hydrogen-carbonate} \rightarrow \text{sodium citrate} + \text{carbon dioxide} + \text{water}$$

Figure 9.2 The sherbet inside these sweets makes your mouth feel cool.

2 What makes the sherbet fizz?

Cooking

When foods are cooked, the heat energy they take in allows chemical reactions to take place which change their structure and taste.

Figure 9.3 Cooking brings about chemical reactions that change food.

Heating limestone

Limestone is made from calcium carbonate. When limestone is heated it breaks down to form calcium oxide and carbon dioxide. Calcium oxide is normally called lime and it is used to make bricks, plaster, glass and paper. When water is added to lime, slaked lime is made which is used to neutralise acid soils, to treat sewage to make it harmless, and in food processing.

Large amounts of limestone are converted into lime in a lime kiln (Figure 9.4). Small limestone rocks are poured into the top of a kiln, which is then sealed. Heat from gas burners decomposes the limestone. Streams of air entering the bottom of the kiln carry away carbon dioxide from the top of the kiln and prevent it reacting with the calcium oxide. If the carbon dioxide did react with the calcium oxide, calcium carbonate would form again.

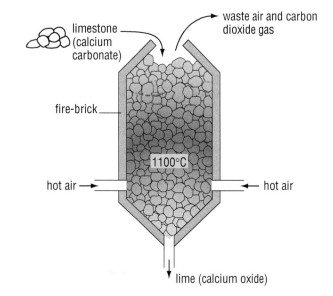

Figure 9.4 A lime kiln

3 Write the word equation for the endothermic reaction that occurs in a lime kiln.

4 If air was not allowed to stream through the kiln, how would the production of lime be affected? Explain your answer.

5 Lime kilns are very large. Why is this necessary?

Photosynthesis

The food used by living things is manufactured by plants. Plants do not release energy as they make food – they take energy in, from sunlight. This food-making process is called photosynthesis (page 10). Carbon dioxide and water are used to trap the energy and make glucose. The word equation for this process is:

carbon dioxide + water → glucose + oxygen

6 Where does the plant get the energy it needs for the chemical reactions in photosynthesis?

There are many chemical reactions which take place in the process of photosynthesis. This equation is just a summary of those reactions.

Exothermic reactions

The exothermic reaction that is most familiar to everyone is **combustion**. When a flame develops in this reaction, combustion is then called burning. We say that a burning substance is on fire.

Burning

Many substances are burned to provide heat or light. They are called **fuels**. Wood, coal, coke, charcoal, oil, diesel oil, petrol, natural gas and wax are all examples of fuels. The heat may be used to warm buildings, cook meals, make chemicals in industry, expand gases in vehicle engines and turn water into steam to drive generators in power stations. Some gases, oils and waxes are burned in lamps to provide light.

Natural gas is an example of a **hydrocarbon**. It is made of carbon and hydrogen. When natural gas burns, carbon dioxide and water (hydrogen oxide) are produced. Many other fuels such as coal, coke and petrol contain hydrocarbons.

Figure 9.5 The burning oil in these lamps gives out light.

7 Give a use for each of the fuels listed in the paragraph on burning. How many different uses can you find?

Investigating a burning candle

A candle can be used to investigate how fuels burn.

● **Investigation 1:** If a burning candle is put under a thistle funnel attached to the apparatus shown in Figure 9.6 and the suction pump is switched on, a liquid collects in the U-tube and the limewater turns cloudy. When the liquid is tested with cobalt chloride paper, the paper turns from blue to pink. This shows that the liquid is water. The cloudiness in the limewater indicates that carbon dioxide has passed into it.

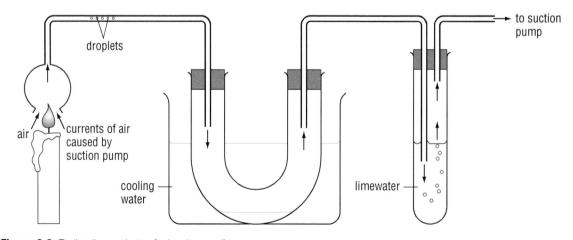

Figure 9.6 Testing the products of a burning candle

● **Investigation 2:** If a beaker is placed over a burning candle, the candle will burn for a while and then go out. A change has taken place in the air that makes it incapable of letting things burn in it.

The test for oxygen is made by plunging a glowing splint of wood into the gas being tested. If the gas is oxygen, the splint bursts into flame. When air from around the burned-out candle is tested for oxygen, the glowing splint goes out. This indicates that oxygen is no longer present. The oxygen in the air under the beaker has been used up by the burning candle.

From the information provided by these two investigations with candles, the following word equation can be set out:

hydrocarbon + oxygen → carbon dioxide + water

Natural gas is a hydrocarbon called methane. When it burns, it breaks down exactly like the hydrocarbons in candle wax. The word equation for this reaction is:

methane + oxygen → carbon dioxide + water

Both of these word equations are examples of complete combustion. This only happens when there is enough oxygen available.

The danger of incomplete combustion

If there is insufficient oxygen to support complete combustion, incomplete combustion takes place. Carbon monoxide is a very dangerous chemical produced by incomplete combustion. It is produced instead of carbon dioxide. Carbon monoxide is produced in car engines and is released in the exhaust fumes.

Incomplete combustion also occurs when a gas fire has been incorrectly fitted and cannot draw enough oxygen from the room it is heating. Carbon monoxide is a colourless, odourless gas so you do not know when it is being produced. If it is breathed in, it stops the blood taking up oxygen and circulating it round the body.

People have died from breathing carbon monoxide from badly fitted gas fires. All gas fires must be fitted by a trained engineer and used in a well-ventilated room so that there is enough air passing through the fire to provide oxygen for complete combustion of the gas.

8 What happens to the carbon in natural gas when the gas burns in a badly fitted gas fire? Explain your answer.

9 Which fuels would you take on a camping trip? Explain your choices.

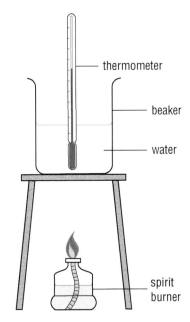

Figure 9.7 A simple calorimeter

10 How could you use the procedure with the simple calorimeter to compare the energy released by different fuels?

Measuring energy in a fuel

The energy in a fuel can be found by collecting the heat energy released from a burning fuel and measuring it. The device used to do this is called a **calorimeter**.

A beaker of water can be used as a very simple calorimeter (Figure 9.7) to measure the energy of a fuel in the following way.

1 Pour an amount of cold water into the beaker and record its temperature.
2 Weigh the spirit burner containing the fuel and record its mass.
3 Light the burner so that the burning fuel heats the water.
4 When the temperature has risen by about 10 °C, put out the burner flame, record the temperature of the water, and then reweigh the burner and fuel.
5 The amount of fuel used is found by subtracting the mass of the burner and fuel after the experiment from the mass of the burner and fuel before the experiment.
6 The rise in temperature is found by subtracting the temperature of the water at the beginning of the experiment from the temperature of the water at the end of the experiment.
7 The energy released by burning the mass of fuel is indicated by the rise in temperature of the water.

At the beginning of the *Cambridge Checkpoint Science* course, in the Introduction to *Student's Book 1*, we saw that scientists sometimes make apparatus for their investigations. In Figure 9.8 you can see how scientists modified the simple equipment shown in Figure 9.7 to make much more accurate measurements of the energy released as a fuel burns.

This calorimeter consists of a chamber in which the fuel is burnt, a water jacket around the chamber to collect the heat, and a system of tubes that bring air to the fuel and carry away the hot gases so that they release their heat into the water jacket.

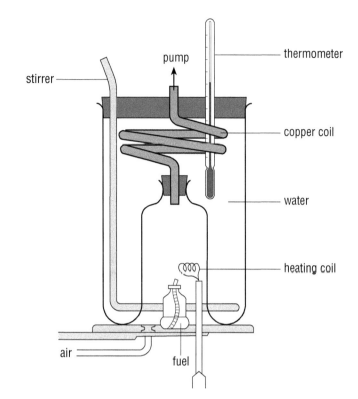

Figure 9.8 A specially designed calorimeter

11 Why is air pumped through the chamber instead of just letting the fuel use the air that is present?

12 Why is the pipe carrying the hot gases made of a copper coil?

13 Why is the water stirred?

14 How do you find the rise in temperature of the water?

15 If a calorimeter had 500 g of water in its jacket and the water temperature rose by 25 °C, how many joules of energy were released by the fuel?

16 If the fuel had a mass of 2 g, how much heat energy does each gram of fuel release?

The mass of the fuel is found before it is placed in the calorimeter. The temperature of the water is taken before the fuel is ignited. Air passes over the fuel as it burns and the water in the jacket is stirred. The temperature of the water is not taken until the fuel burns out.

The data obtained by using the calorimeter is used in calculations with other evidence in the following way.

It has been found that 4.2 kilojoules of energy raises the temperature of 1000 g of water by 1 °C. If the volume of water in the calorimeter is 500 g, 2.1 kJ will raise the temperature by 1 °C. The amount of energy released by the fuel is found by multiplying the rise in temperature by 2.1.

An alternative way to find the energy released by a fuel is to repeat the experiment with a wire heating element in the chamber. Electric current is passed through a joulemeter and through the element, and the water temperature is allowed to rise until it reaches the same temperature as that when the fuel burnt out. The current is then switched off and the joulemeter is read to find the amount of energy required to cause this temperature rise.

The energy in different fuels has been measured and some results are shown in Table 9.1.

17 How many grams of biscuit (carbohydrates) would you need to provide the same amount of energy as 100 g of coke?

18 Bjorn and Ingrid live in a country with cold winters. They live in identical houses but Bjorn heats his house with coal while Ingrid heats her house with wood. They both fill up their fuel stores for the winter. Who will need the larger store? Explain your answer.

Table 9.1 Energy in various fuels

Fuel	Energy/kJ/g
natural gas	56
oil	48
coal	34
coke	32
wood	22
carbohydrates (e.g. starch and sugars)	16

Improving efficiency

Wood was the first fuel. In many parts of the world it is still used as a fuel today. It is often in short supply so ways have been developed to use the fuel more efficiently. Figure 9.9 shows a stove that has been developed in Sri Lanka to provide heat for cooking two pots for a meal at once. One pot is used to boil rice while the other is used to cook vegetables.

The damper can be raised or lowered and used to control the amount of air reaching the fire. This in turn controls the burning of the wood. The baffle is used to control the direction of the flames. They can be made to go straight up and heat the pot above them.

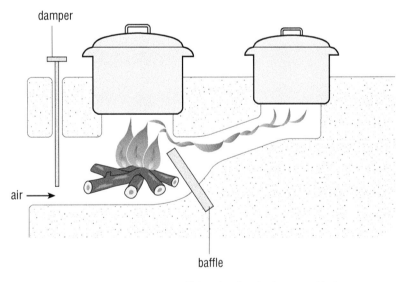

19 How will designing more efficient stoves help conserve fuel?
20 How will using the damper make the wood burn more slowly or more quickly?
21 In the past people used to put a pan on three stones over a fire. Why is the stove in Figure 9.9 an improvement?

Figure 9.9 Heating two pots is more efficient than heating just one with the same amount of fuel.

Respiration

All life processes require energy. Living things obtain their energy from their food. The process in which energy is released from food is called **respiration**. In this process a food substance called glucose reacts with oxygen. The word equation for this reaction is:

glucose + oxygen → carbon dioxide + water

The energy released in respiration is used to make substances inside the body and to make a body move. Some of the energy is released as heat (Figure 9.10).

Figure 9.10 This runner has been generating a great deal of heat.

22 Is food a fuel? Explain your answer.

23 Is respiration like burning? Explain your answer.

Oxidation

When a substance burns in air it reacts with oxygen. This is called an **oxidation** reaction. Oxygen combines with elements in the substance to form a compound. In the examples in this chapter, incomplete combustion causes carbon in a fuel to be oxidised to carbon monoxide. Complete combustion results in carbon being oxidised to carbon dioxide. Oxidation also occurs in respiration where the carbon in glucose is oxidised to carbon dioxide.

Rusting is another example of oxidation (Figure 10.1b, page 145). When water vapour from the air condenses on iron or steel it forms a film on the surface of the metal. Oxygen dissolves in the water and reacts with the metal to form iron oxide. This forms brown flakes of rust, which break off from the surface and expose more metal to the oxygen dissolved in the water. The iron or steel continues to produce rust until it has completely corroded. The word equation for the reaction is:

iron + oxygen → iron oxide

Hand warmers and rusting

Rusting is an exothermic reaction but as the reaction is slow the heat is produced in small amounts and quickly spreads into the air so that a rusting object does not feel warm. Useful heat can be produced from rusting in the design of some kinds of hand warmers. Iron powder is mixed with particles of charcoal (carbon), salt water and an insulating substance such as vermiculite. All these constituents of the hand warmer are enclosed in a sealed airtight package. When the hand warmer is required the seal is broken and air mixes with the constituents, so that an oxidation reaction takes place between the iron powder and oxygen. The salt water acts as a catalyst (page 164) and speeds up the reaction so a large quantity of heat can be produced quickly. The carbon particles take up the heat and spread it out through the package, while the insulating material prevents it escaping too quickly so that a smaller amount of heat is released steadily to warm the hands.

◆ SUMMARY ◆

◆ Melting is an endothermic reaction (*see page 135*).
◆ The chemicals in sherbet take part in an endothermic reaction (*see page 136*).
◆ The cooking of food is an endothermic reaction (*see page 136*).
◆ Limestone breaks down in an endothermic reaction (*see page 137*).
◆ Photosynthesis is an endothermic reaction (*see page 137*).
◆ Burning is an exothermic reaction (*see page 138*).
◆ Incomplete combustion is dangerous (*see page 139*).
◆ The energy in a fuel can be measured (*see page 140*).
◆ Respiration is an exothermic reaction (*see page 143*).
◆ Burning and rusting are examples of oxidation reactions (*see page 143*).

End of chapter questions

1 Plan an investigation to compare the amounts of energy in two fuels.
2 What kind of reaction, exothermic or endothermic, takes place when:
   a) a plant makes food in sunlight
   b) wood is used to cook food
   c) the food in the plant is cooked on the fire
   d) some of the food is used in respiration in your body cells?

(10) Patterns of reactivity

◆ How metals react with oxygen
◆ How metals react with water
◆ How metals react with acids
◆ Displacement reactions
◆ The reactivity series

Sodium is so reactive that it has to be kept under a layer of oil. The reason for this is that it will readily react with water and oxygen in the air and produce flames. Sodium is a very soft metal and can be cut with a knife. If a small piece is cut and exposed to the air, its shiny surface quickly tarnishes as it reacts with the air.

On page 143, we saw that iron reacts with damp air and turns into rust. This reaction takes place much more slowly than the reaction between sodium and the air and suggests that metals have different speeds at which they react with other chemicals.

Figure 10.1 Sodium reacts with oxygen so quickly that it must be stored under oil to prevent it bursting into flames (left). Iron reacts more slowly, but over time the results can be dramatic (right).

Reaction of metals with oxygen

Here are some descriptions of the reactions that take place when certain metals are heated with oxygen.

- Copper develops a covering of a black powder without glowing or bursting into flame.
- Iron glows and produces yellow sparks; a black powder is left behind.
- Sodium only needs a little heat to make it burst into yellow flames and burn quickly to leave a white powder behind (Figure 10.2).
- Gold is not changed after it has been heated and then left to cool.

1 Arrange the metals mentioned in this section in order of their reactivity with oxygen. Start with what you consider to be the most reactive metal.

Figure 10.2 Sodium burns rapidly in a gas jar of oxygen (left). Sodium oxide powder is left behind (right).

Reaction of metals with water

Here are some descriptions of the reactions that take place between water and metals. In the study of these reactions, the metals were first tested with cold water. If there was no reaction, the test was repeated with steam using the apparatus shown in Figure 10.3.

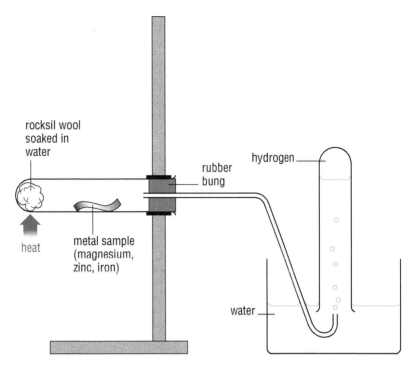

Figure 10.3 This apparatus can be used to investigate the reaction of a metal with water (in the form of steam).

- Calcium sinks in cold water and bubbles of hydrogen form on its surface, slowly at first. The bubbles then increase in number quickly and the water becomes cloudy as calcium hydroxide forms. The bubbles of gas can be collected by placing a test tube filled with water over the fizzing metal. The gas pushes the water out of the test tube. If the tube, now filled with gas, is quickly raised out of the water and a lighted splint held beneath its mouth, a popping sound is heard. The hydrogen in the tube combines with oxygen in the air and this explosive reaction makes the popping sound.
- Copper sinks in cold water and does not react with it. It also does not react with steam.
- Sodium floats on the surface of the water and a fizzing sound is heard as bubbles of hydrogen are quickly produced around it. The production of the gas may push the metal across the water surface and against the side of the container, where the metal bursts into flame. A clear solution of sodium hydroxide forms.
- Iron sinks in water and no bubbles of hydrogen form. When the metal is heated in steam, hydrogen is produced slowly.

2 Arrange the metals in this section in order of their reactivity with water.

3 Which metals would not be put in the apparatus in Figure 10.3 to see if they reacted with steam? Explain your answer.

4 Which metals are less dense than water? (See Chapter 13 to find out about density.)

5 Water is a compound of hydrogen and oxygen and could be called hydrogen oxide. When hydrogen is released as a metal reacts with steam, what do you think is the other product of the reaction?

6 In the home, copper is used for the hot water tank, while steel (a modified form of iron) is used to make the cold water tank. Why can steel not be used to make the hot water tank?

7 Based on the way sodium and potassium react with water, as described in the last section, which metal do you think will react more violently with acids?

● Magnesium sinks in water. Bubbles of hydrogen are produced only very slowly and a solution of magnesium hydroxide is formed. When the metal is heated in steam hydrogen is produced quickly.

● Potassium floats on water and bursts into flames immediately (Figure 10.4). Hydrogen bubbles are rapidly produced around the metal. A clear solution of potassium hydroxide forms.

Figure 10.4 The reaction of potassium with water gives out enough heat to set alight the hydrogen produced, which burns with a pink flame because potassium is present. The heat also melts the potassium metal, which forms a molten ball that floats on the water's surface.

Reaction of metals with acids

Sodium and potassium react violently with acids. Other metals react less violently and their reactions with dilute hydrochloric acid can be studied using a test tube as shown in Figure 10.5 or with the set up of apparatus shown in Figure 10.6.

A piece of metal is placed in a test tube or flask, then acid is poured in and the bubbles of hydrogen gas are observed. The reactivity of metals can be compared by simply comparing the amount of hydrogen gas produced in the reaction between the acid and different metals. The general word equation for this reaction is:

metal + hydrochloric acid → metal chloride + hydrogen

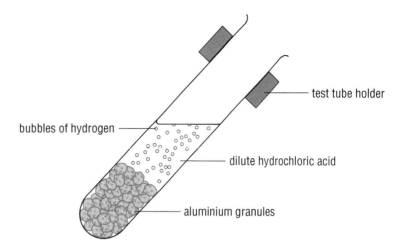

bubbles of hydrogen

test tube holder

dilute hydrochloric acid

aluminium granules

Figure 10.5 Aluminium reacts with dilute hydrochloric acid, producing bubbles of hydrogen gas and aluminium chloride solution.

Figure 10.6 Apparatus used in investigating the reaction of some metals (not sodium or potassium) and hydrochloric acid.

The apparatus in Figure 10.6 is designed to collect any hydrogen that is produced in the reaction. When the acid is poured in down the thistle funnel it should reach a level above the bottom of the funnel tube. The hydrogen escapes from bubbles at the surface of the solution and pushes air out of the flask and down the delivery tube. A minute should be allowed for the air to escape from the end of the delivery tube, and then a test tube full of water can be put over it to collect the hydrogen. As the gas collects in the upturned test tube it pushes water out at the bottom. When the tube is full of gas it can be closed with a bung or stopper and replaced with another tube to collect any more gas that is produced.

8 What might happen to the hydrogen if the bottom of the funnel was above the surface of the liquid?

9 What is the word equation for the reaction of each of the following metals with hydrochloric acid?
   a) magnesium
   b) iron
   c) zinc

10 Why was a concentrated solution used if there was no reaction with the dilute solution?

11 Arrange the metals in this section in order of their reactivity with hydrochloric acid.

12 If a metal that had reacted very slowly with dilute acid was tested with a concentrated one, what would you predict would happen?

Here are some descriptions of the reactions that take place between different metals and hydrochloric acid. The metals were first tested with dilute hydrochloric acid. If the reaction did not take place, they were tested with concentrated hydrochloric acid.

- Lead did not react with dilute hydrochloric acid but when tested with concentrated acid, bubbles of hydrogen gas were produced slowly.
- Zinc reacted quite slowly with dilute hydrochloric acid to produce bubbles of hydrogen.
- Copper did not react with either dilute or concentrated hydrochloric acid.
- Magnesium reacted quickly with dilute hydrochloric acid to produce bubbles of hydrogen.

Displacement reactions

When metals react with acids, they **displace** hydrogen from the acid and form a salt solution. In a similar way, a more reactive metal can displace a less reactive metal from a salt solution of the metal.

When a copper wire is suspended in a solution of silver sulfate, the copper dissolves into the solution to form copper sulfate and silver metal comes out of the solution and settles on the wire (Figure 10.7).

Figure 10.7 Copper wire coils in silver sulfate solution. Silver is formed on the wire.

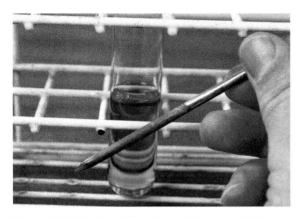

Figure 10.8 This iron nail has been left in copper sulfate solution. Copper has formed on the nail.

If an iron nail is placed in copper sulfate solution, the iron dissolves to form a pale green iron sulfate solution and the copper comes out of the solution and coats the nail (Figure 10.8).

13 From the information about these two displacement reactions, arrange the three metals (copper, silver and iron) in order of reactivity – starting with the most reactive.

14 Look at Table 10.1. How do the reactions of zinc with oxygen, water and acid compare with those of iron with these substances?

15 From Table 10.1, in displacement reactions would you expect zinc to displace each of the following metals? Explain your answer.
  a) iron
  b) sodium
  c) gold

The reactivity series

In Chapter 8 we saw how chemists sorted out the elements into order according to their atomic number and their properties to produce the periodic table. Chemists have also looked at the reactivity of metals and arranged them in order too. The metals are arranged in order of their reactivity, starting with the most reactive. The **reactivity series** is produced by studying the reactions of metals with oxygen, water, hydrochloric acid and solutions of metal salts. Table 10.1 shows eleven metals in the reactivity series and summarises their reactions with oxygen, water and hydrochloric acid.

Table 10.1 The reactivity series

Metal	Reaction with oxygen	Reaction with water	Reaction with acid
potassium	oxide forms very vigorously	produces hydrogen with cold water	violent reaction
sodium			
calcium		produces hydrogen with steam	rate of reaction decreases down the table
magnesium			
aluminium			
zinc			
iron	oxide forms slowly		
tin	oxide forms without burning	no reaction with water or steam	very slow reaction
copper			no reaction
silver	no reaction		
gold			

◆ SUMMARY ◆

◆ Metals react at different speeds with oxygen (*see page 146*).
◆ Metals react at different speeds with water (*see page 146*).
◆ Metals react at different speeds with acids (*see page 148*).
◆ A more reactive metal can displace a less reactive metal from its salt (*see page 150*).
◆ Metals can be arranged in order of their reactivity. This is called the reactivity series (*see page 151*).

End of chapter questions

1 Figure 10.9 shows the apparatus and materials that are required for a thermit (or thermite) reaction to take place. The heat generated in this reaction can be used to cut metal or weld it together. It can be used to weld railway lines together while they are in place on the ground. The reaction is started by setting the magnesium ribbon on fire and when the heat reaches the mixture of aluminium powder and iron oxide a displacement reaction takes place. The magnesium takes no part in this reaction. A great deal of heat is given out, so that the metal produced is in molten form. When welding railway lines, this molten metal is allowed to flow into a mould around the break, where it cools and becomes solid, joining the two parts together.

a) What do you think takes place in this displacement reaction?
b) Write a word equation for the reaction.
c) Explain why the reaction takes place. Refer to the reactivity series in your answer.

2 Why do you think you can find silver and gold on their own in rocks but calcium and magnesium are combined with other elements to make compounds such as calcium carbonate and magnesium sulfate? Refer to the reactivity series in your answer.

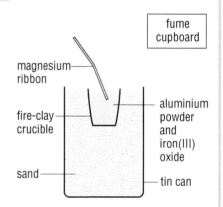

Figure 10.9 This apparatus can be used to carry out the thermit reaction in the laboratory.

♦ Word equations for making salts
♦ Acids and their salts
♦ Preparing a salt from a metal and an acid
♦ Preparing a salt from a metal carbonate and an acid

Some people have difficulty thinking about the word '**salt**'. All they can think about is sodium chloride – common salt. There are many different salts and some of them have a wide range of uses.

Calcium chloride is a salt that forms white crystals. It is used to absorb moisture from the air and is known as a drying agent. It is also used in food processing, in medicine and in speeding up the setting of concrete.

Zinc sulfate also forms white crystals. It is used in making cosmetics, some deodorants, the textile material rayon and glue, and in the bleaching of paper. It is also used as a herbicide and in the treatment of sewage, and it is used by chemists in investigations of substances to find their chemical components.

Figure 11.1 Zinc sulfate is a metal salt that has many uses in industry, including the manufacture of rayon fibres from cellulose.

1 Do you think all metal salts can be prepared by adding the metal to an acid? Explain your answer.

In this chapter we are going to look at two methods of preparing salts – using metals and acids, and using metal carbonates and acids. The chemical reactions that take place in these two methods of preparation are represented by these general word equations:

acid + metal → salt + hydrogen

acid + carbonate → salt + carbon dioxide + water

Acids and their salts

The three acids that can be used to make salts are hydrochloric acid, sulfuric acid and nitric acid. The salts produced by these acids are chlorides, sulfates and nitrates respectively.

Figure 11.2 Some examples of chloride, sulfate and nitrate salts

Preparing a salt from a metal and an acid
Example – preparation of zinc chloride

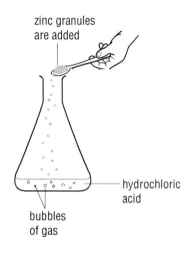

Figure 11.3 Adding zinc to hydrochloric acid

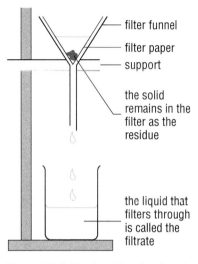

filter funnel

filter paper

support

the solid remains in the filter as the residue

the liquid that filters through is called the filtrate

Figure 11.4 Filtration with a filter funnel

Figure 11.5 Evaporation

Granulated zinc is added to hydrochloric acid in a flask (Figure 11.3). Bubbles of gas rise from the metal, pass through the liquid and escape into the air. Eventually the bubbles are no longer produced and some metal remains in the flask. The contents of the flask are poured into filter paper in a filter funnel (Figure 11.4). The zinc metal remains behind and the liquid passes through and falls into a beaker. The liquid is poured into an evaporating dish and heated gently until some solid appears (Figure 11.5). The mixture is then left to cool and more evaporation takes place. After this the mixture is filtered again.

2 Why do you think that granulated zinc is used instead of a block? You may have to look at page 161 to help you answer.

3 What passed through the filter paper when the flask was emptied?

4 Why was the solution heated before it was left in an evaporating dish?

5 Write the word equation for this reaction.

6 Zinc sulfate can be prepared in a similar way. Write a word equation for the reaction.

7 Write the word equations for the reactions between the following metals and acids.
  a) magnesium and sulfuric acid
  b) iron and nitric acid
  c) calcium and hydrochloric acid
  d) lead and sulfuric acid
  e) aluminium and hydrochloric acid
  f) tin and nitric acid

Preparing a salt from a metal carbonate and an acid

Example – preparation of calcium chloride

The carbonate used in this reaction is calcium carbonate in the form of marble chips. Some marble chips are added to hydrochloric acid in a flask. Bubbles are produced and the chips dissolve. Some more chips are added and more bubbles are produced and then the reaction stops and some of the chips are left in the solution. The contents of the flask are poured into a filter paper in a filter funnel and the solution and chips are separated (Figure 11.6). The liquid is poured into an evaporating dish and heated until some solid appears. The mixture is then left to cool and more evaporation takes place. When the mixture has been left to cool it is filtered again.

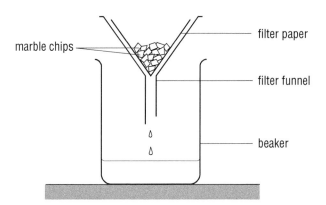

Figure 11.6 Filtering out marble chips using a filter funnel

8 Why did all the chips added at first dissolve?

9 Why did some of the chips added later not dissolve?

10 Why were the contents of the flask filtered?

11 Write the word equation for the reaction.

12 Write the word equations for the reactions between the following metal carbonates and acids.
a) zinc carbonate and sulfuric acid
b) aluminium carbonate and hydrochloric acid
c) magnesium carbonate and nitric acid
d) copper carbonate and sulfuric acid
e) calcium carbonate and nitric acid
f) lead carbonate and hydrochloric acid

Salimuzzaman Siddiqui and chemicals from natural products

The activities involved in preparing salts are used on a daily basis by chemists all over the world. Some of this work is directed at extracting chemicals from plants for use in medicines and insecticides. One of the pioneers in this work was Salimuzzaman Siddiqui (1897–1994) who was born in Subeha in India and then went on to work in Pakistan.

Salimuzzaman Siddiqui attended school in Lucknow and had many interests including literature, music and calligraphy. He went on to university in Aligarh, where he graduated in philosophy and the Persian language. He moved to London to study medicine but after a year he moved to Frankfurt in Germany to study chemistry. While there he continued his interest in art and literature. He held the first international exhibition of his paintings in the city and translated the work of a German poet into Urdu, which is still popular today.

Siddiqui returned to India and, guided by Hakim Ajmal Khan, he set up a research institute. A plant known as snake root or sarpagandha (*Rauwolfia serpentina*) has been used for thousands of years in India to treat ailments and Siddiqui began work on extracting chemicals from its roots. In 1931, he extracted from these roots a substance that cures abnormal beating of the heart. He named it Ajmaline in honour of Hakim Ajmal Khan who helped him on his return to India. Siddiqui went on to discover more chemicals from the plant's roots and many are still used today to treat heart disease and mental disorders.

In 1942, he used a range of solvents including ether and dilute alcohol to extract chemicals from the neem tree or 'heal-all' (*Azadirachta indica*). These chemicals can be used to kill bacteria, viruses, fungi, parasitic worms and as natural insecticides. He continued with his work, discovering over 50 substances, and went on to discover more in research projects with other scientists and their students.

In 1951, Siddiqui was asked to set up research programmes in Pakistan and in 1953 he founded the Pakistan Academy of Sciences and the Pakistan Council of Scientific and Industrial Research. In 1967, he set up the Post-Graduate Institute of Chemistry at the University of Karachi, which in 1976 was renamed the Hussain Ebrahim Jamal Research Institute of Chemistry. Institutes and academies are places where scientists can gather together to discuss their work, and Siddiqui built up their work in chemistry and natural products to international importance. In 1983, he helped to set up, and became a founder member of, the Third World Academy of Sciences. At the age of 93 he retired from running the institute but carried on research in his laboratory.

Figure A Siddiqui discovered many extremely useful chemical substances in the tissues of plants such as this neem tree.

1 Would you say that the young Salimuzzaman Siddiqui was interested in the world around him? Explain your answer.
2 What signs of a scientist must he have displayed in his early interests?
3 Why do you think the scientific names of the plants have been included in the text?
4 A scientist should be able to work with others. What evidence can you see of Siddiqui working with others?
5 Would you say that Siddiqui had a life-long passion for science? Explain your answer.

6 What evidence do you think Siddiqui might have considered when selecting the sarpagandha and neem tree for his research?
7 Where do you think Siddiqui might have obtained his previous knowledge to use solvents such as ether and alcohol?
8 How do you think Siddiqui helped other scientists to explain and communicate their results to others?

◆ SUMMARY ◆

◆ The preparation of many salts can be summarised by two word equations (*see page 154*).

◆ Three acids used in making salts are hydrochloric acid, sulfuric acid and nitric acid (*see page 154*).

◆ Zinc chloride is prepared from granulated zinc and hydrochloric acid (*see page 155*).

◆ Calcium chloride is prepared from marble chips and hydrochloric acid (*see page 156*).

◆ The life of Salimuzzaman Siddiqui shows how scientists worked in the past, using evidence, exhibiting creative thought and performing experiments (*see page 157*).

End of chapter questions

1 You are given a metal and asked to make its metal chloride. Write down a plan to carry out the experiment.

2 You are given a metal carbonate and asked to make its metal sulfate. How will you modify your plan in answer to question **1** to make the salt?

3 You are asked to try to identify the metal in the salt you have made and are given this table to help you.

Element	Colour of flame
aluminium	silvery-white
barium	apple green
calcium	orange
caesium	blue
copper	green
lithium	red
magnesium	white
sodium	golden yellow

a) To what does the table refer?

b) How might you plan an investigation, using the table to help you, to identify salts of aluminium, calcium, copper and magnesium?

12 Rates of reaction

◆ Measuring rate of reaction by the change in mass of the reactants
◆ Measuring rate of reaction by the change in volume of the product
◆ The effect of concentration
◆ The effect of particle size
◆ The effect of temperature
◆ Catalysts

Chemical reactions take place at different **rates**. The chemical reaction that blows rock out of a cliff face in a quarry is a very fast reaction. Some reactions, such as the setting of concrete, are very slow and may take two days or more to complete.

Figure 12.1 An explosion in a quarry happens as a result of a very rapid reaction.

Figure 12.2 When concrete sets, the reaction takes two days or longer.

Measuring rate of reaction

In chemistry, the rate of a reaction is studied by considering the rate at which the chemicals in the reaction change. Rate is a measure of the change in a certain amount of time. The rate may show how much the mass of the reactants changes in a certain amount of time, or how much product is produced in a certain amount of time.

Change in mass of reactants

After the mass of the reactants (marble chips and hydrochloric acid) has been recorded, the reactants are mixed together in the flask and their joint mass is recorded (Figure 12.3). The mass is then recorded regularly over a certain number of minutes. The word equation for the reaction is:

$$\text{calcium carbonate} + \text{hydrochloric acid} \rightarrow \text{calcium chloride} + \text{carbon dioxide} + \text{water}$$

The carbon dioxide escapes from the flask and accounts for the change in mass.

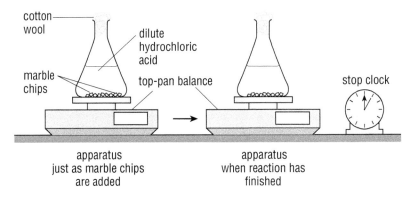

Figure 12.3 Finding the mass of reactants before, during and after the reaction

Change in volume of a product

In Figure 12.4, when the reactants (magnesium and hydrochloric acid) are mixed together, hydrogen is produced, as the word equation describes:

$$\text{magnesium} + \text{hydrochloric acid} \rightarrow \text{magnesium chloride} + \text{hydrogen}$$

1 Why do you think the production of hydrogen is estimated in a different way to the production of carbon dioxide?

As the gas is produced, it pushes the plunger in the syringe to the right and the volume produced every minute can be measured.

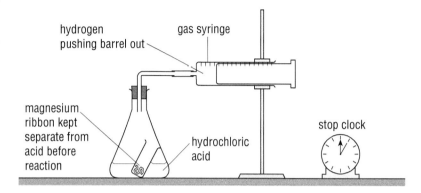

Figure 12.4 The volume of hydrogen produced can be measured.

Factors affecting rates of reaction

Concentration

The concentration of a solution is a measure of the amount of solute dissolved in it. A solution of low concentration has a small amount of solute dissolved in it. A solution of high concentration has a large amount of solute dissolved in it. If the concentration of a reactant is increased, the rate of reaction increases. If the concentration of one reactant is doubled the rate of the reaction may be doubled.

Particle size

If a piece of coal is heated, it produces a flame and burns steadily in air. If coal dust is heated, it explodes. The reason for this difference in reaction rate is due to the surface area of coal in contact with oxygen in the air. When a piece of coal is ground into dust, it has a much larger surface area in contact with the oxygen in the air. This suggests that particle size affects the rate of reaction.

When a solid, such as a piece of coal, breaks up into smaller particles its surface area increases, as the following example shows. Imagine a cube-shaped piece of coal with sides 2 cm long (Figure 12.5 on page 162). It has six surfaces. Each one is $2 \times 2 = 4\,cm^2$ in area. The surface area of the cube is $6 \times 4 = 24\,cm^2$.

2 In a model volcano, some vinegar is poured onto sodium hydrogencarbonate. A reaction takes place producing a fizzy liquid that flows down the sides of the volcano like lava. If water is added to the vinegar and the reaction is repeated, will the 'eruption' of the volcano be stronger or weaker? Explain your answer.

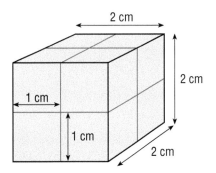

Figure 12.5 One large cube can be cut into eight smaller cubes.

If the cube is broken into eight cubes, each with a side of 1 cm, the surface area of each cube would be $6 \times 1 \times 1 = 6\,cm^2$. As there are now eight cubes, their total surface area is $8 \times 6 = 48\,cm^2$. The surface area has doubled. If the eight cubes were broken up into smaller pieces, the surface area would increase even more.

The surface area of a reactant is its point of contact with other reactants. If the surface area is increased, the reactants can come into contact more rapidly and the reaction rate will increase.

3 The graph in Figure 12.6 shows the volume of carbon dioxide produced when large marble chips take part in a reaction with hydrochloric acid.
  a) Make a copy of the graph and draw in the line you would expect to see when smaller chips are used.
  b) Explain your answer.

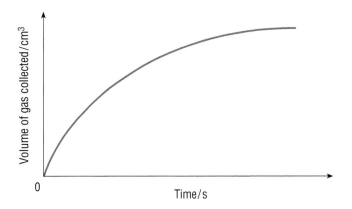

Figure 12.6 A graph showing the volume of carbon dioxide produced over time

Temperature

The rate of reaction increases if the temperature is raised. The rate decreases if the temperature is lowered. If the temperature of the reactants is raised by 10°C the speed of the reaction may be doubled.

4 Two cartons of milk were opened and one was left by a radiator while the other was placed in a fridge. How will the milk in the two cartons differ after three days? Explain your answer.

Measuring the effect of temperature

Sodium thiosulfate is a substance that dissolves in water. When hydrochloric acid is added to a solution of sodium thiosulfate, sodium chloride, water, sulfur and sulfur dioxide are produced. The sulfur forms a precipitate that clouds the solution and the speed at which this cloudiness appears can be used as a measure of the rate of the reaction. An investigation can be carried out in the following way.

1 A flask containing the reactants is placed over a piece of paper with a cross on it.

2 The reactants are viewed from the top of the flask and a stop clock is started.

3 When the sulfur precipitate clouds the solution so much that the cross can no longer be seen, the stop clock is stopped and the time is recorded (Figure 12.7).

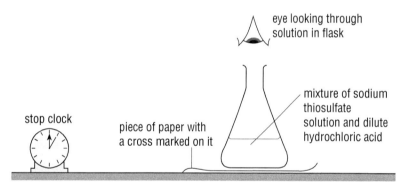

Figure 12.7 An experiment to assess how temperature alters rate of reaction

5 a) Plot the data in Table 12.1 as a graph.

b) What is the shape of the graph?

c) What would you predict the time of the reaction to be at 32.5 °C?

d) What would you predict the temperature to be if the reaction took 76 seconds?

6 Write a word equation for the reaction between sodium thiosulfate and hydrochloric acid.

If the experiment is repeated several times with the reactants at increasingly higher temperatures, a table of data can be produced as Table 12.1 shows.

Table 12.1 Reaction time at different temperatures

Temperature/°C	Time of reaction/s
25	110
30	80
35	60
40	46
45	38
50	30

Catalysts

A **catalyst** is a substance that is added to the reactants to increase the rate at which they react. At the end of the reaction the catalyst remains unchanged chemically. This means that it can be used again. Only a small amount of catalyst is needed to produce a large increase in the rate of a reaction. A catalyst usually only works at speeding up one reaction. It does not speed up a range of reactions.

Using a catalyst in the laboratory

Hydrogen peroxide is a liquid that very slowly breaks down into water and oxygen. However, the rate of reaction can be greatly increased by adding a small amount of manganese dioxide (Figure 12.8). The reaction becomes so fast that the liquid fizzes as the oxygen escapes.

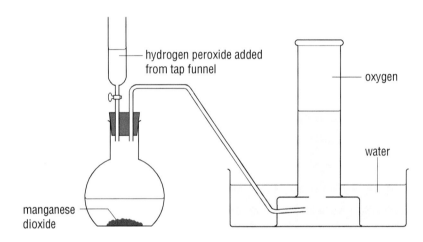

Figure 12.8 Hydrogen peroxide breaks down more quickly when manganese dioxide is added to it.

Catalysts and air pollution

Car engines produce a range of gases. Some of them such as carbon monoxide, nitrous oxides and hydrocarbons are harmful. The catalytic converter used on cars is a device that forms part of the exhaust system. Inside the converter is a catalyst made of platinum and rhodium. The waste gases from the engine take part in chemical reactions in the converter, which produce water, nitrogen and carbon dioxide.

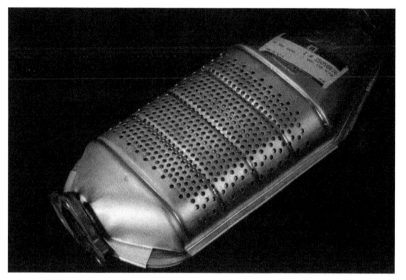

7 The middle of a catalytic converter has a honeycomb structure. Why is this structure used?

Figure 12.9 A catalytic converter removes harmful substances from the exhaust gases of cars by speeding up their reactions to form other products.

Biological catalysts

Biological catalysts are called **enzymes**. They speed up the rates of reaction of life processes in plants and animals. Many chemical reactions take place in the liver and enzymes are present to speed them up. If a piece of liver is placed in hydrogen peroxide it speeds up its decomposition in a similar way to manganese dioxide.

Enzymes are made from proteins. These substances are destroyed by heat. The graph in Figure 12.10 shows how the rate of a reaction catalysed by an enzyme varies with temperature.

The rate of a reaction catalysed by an enzyme is also affected by the pH of the liquid it is in (Figure 12.11).

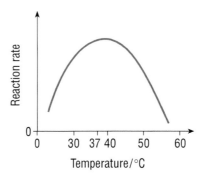

Figure 12.10 Graph showing how reaction rate varies with temperature

8 In Figure 12.10, why does the rate of reaction increase with temperature on the left-hand side of the graph?

9 In Figure 12.10, why does the rate of reaction decrease on the right-hand side of the graph?

10 In Figure 12.11, at what pH is the rate of reaction greatest?

11 Make a drawing of the graph in Figure 12.11 and add to it lines representing:
  a) the rate of a reaction catalysed by an enzyme that works best in acidic conditions
  b) the rate of a reaction catalysed by an enzyme that works best in alkaline conditions.

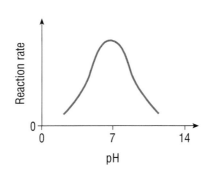

Figure 12.11 Graph showing how reation rate varies with pH

In the human digestive system there are enzymes to break up proteins, fats and carbohydrates in food. Proteins and fats also form most of the stains on dirty clothing and biological washing powders containing enzymes have been developed to break them down. Biological washing powders are not suitable for everyone as some people are allergic to the enzymes and develop a rash when they come into contact with them on clothing.

The particle theory and rates of reaction

We have seen that the particle theory of matter can be used to explain physical changes such as melting and freezing. The particle theory of matter can also be used to explain the factors that affect the rates of reactions. Particles take part in reactions when they collide together so any factors that increase the chance of collisions will increase the rate of reaction.

Concentration

A concentrated solution has more particles in it that are available to react than a dilute solution. This means that increasing the concentration of a solution increases the number of particles and increases the number of collisions and the rate of reaction.

Particle size

Here the word 'particle' does not mean tiny particles as described by the particle theory of matter, but particles like dust and flour. On page 162, we saw how a large cube had a smaller surface area than many smaller ones. Reactions take place at surfaces – so the smaller the surface area the smaller the chance of collisions between the particles in the surface and the particles of the reactant in the liquid or gas next to the surface. Increasing the surface area increases the chance of collisions and so increases the rate of reaction.

Temperature

The speed at which particles move depends on their temperature. If the temperature is raised the speed of the particles increases and they make harder collisions which are more likely to result in reactions. The increased speed of the particles also increases the chance of collisions taking place.

Catalysts

Catalysts have a surface on which reactants can settle and join together. The catalyst increases the chance of the particles meeting and so increases the rate of reaction. Once the particles have met and joined together, they move away from the surface of the catalyst. This leaves room for other particles to join together.

◆ SUMMARY ◆

◆ The rate of reaction is a measure of the rate at which the chemicals in a reaction change (*see page 160*).

◆ Rates of reactions can be found by measuring the changes in the mass of the reactants (*see page 160*).

◆ Rates of reactions can be found by measuring changes in the volume of a product (*see page 160*).

◆ The rate of reaction is affected by the concentration of one reactant (*see page 161*).

◆ The rate of reaction is affected by the particle size of one reactant (*see page 161*).

◆ The rate of reaction is affected by the temperature of the reactants (*see page 162*).

◆ A catalyst is a substance that speeds up the rate of a reaction (*see page 164*).

◆ Biological catalysts are called enzymes (*see page 165*).

End of chapter questions

1 Shazia has built a campfire and it is burning well. Robert collects some damp logs and puts them on the fire. Shazia is annoyed with Robert because the fire now burns more slowly. Why do you think that there has been a change in the rate of reaction?

2 Once the wood has dried out and the fire is burning well again, Robert challenges Shazia to devise a simple experiment using the wood from a dead tree and bush to show that particle size affects the rate of reaction. What do you think Shazia did?

PHYSICS

◆ The density of some common materials
◆ Measuring the density of a rectangular block
◆ Measuring the density of a liquid
◆ The density of gases

Which is heavier – the wood in the trunk of a tree or the metal in a coin? Your first answer might be to say the trunk of the tree, but it can float on water while a coin would quickly sink to the bottom.

To be a fair comparison we need to find the masses of equal volumes. If we found the mass of a piece of wood the size of a coin we would see it was lighter.

Comparing densities

The **density** of a substance is a measure of the amount of matter that is present in a certain volume of it. The following equation shows how the density of a substance can be calculated:

$$\text{density} = \frac{\text{mass}}{\text{volume}}$$

Figure 13.1 Timber can be transported by water because wood is less dense than water, so it floats at the surface.

The basic SI unit of density is found by dividing the unit of mass by the unit of volume, so it is kg/m^3. This is pronounced 'kilograms per metre cubed'. Table 13.1 shows the densities of some solid materials.

In the school laboratory, when small amounts of materials are used, the density of a substance is often calculated using masses measured in grams and volumes in centimetres cubed, giving a density value in g/cm^3. The density value in units of g/cm^3 can be converted to a value in kg/m^3 by multiplying it by 1000. For example, ice was found to have a density of $0.920\,g/cm^3$. This can also be expressed as $0.920 \times 1000 = 920\,kg/m^3$.

1 Arrange the materials in Table 13.1 in order of density, starting with the least dense material.

2 Which is heavier, 1 m³ of steel or 1 m³ of aluminium?

3 Which is heavier, 1 kg of steel or 1 kg of cork?

Table 13.1 The density of some common solid materials

Material	Density/kg/m³
ice	920
cork	250
wood	650
steel	7900
aluminium	2700
copper	8940
lead	11350
gold	19320
polythene	920
perspex	1200
expanded polystyrene	15

Measuring the density of a rectangular solid block

The mass of the block is found by placing the block on a balance (check the balance reads zero first) and reading the scale. The mass in grams is recorded. The volume is found by multiplying the length, width and height of the block together and recording the value in centimetres cubed. The density of the material in the block is found by dividing the mass by the volume and expressing the quantity in the unit g/cm³.

4 A block of material is 8 cm long, 2 cm wide and 3 cm high, and has a mass of 46 g. What is its density?

5 a) Convert the value you found for the density in question 4 to kg/m³.

b) Compare the density of the material in the block with those in Table 13.1. Which materials in the table have densities closest to that of the block?

c) How could you convert the value of a density given in kg/m³ to g/cm³?

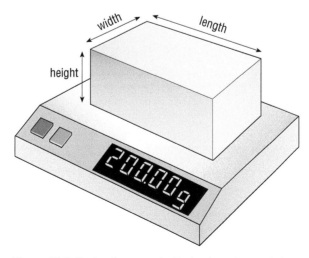

Figure 13.2 Finding the mass of a block using a top-pan balance

Measuring the density of an irregularly shaped solid

The density of an irregularly shaped solid, such as a pebble, can be found in the following way.

The mass of the pebble is found by placing it on a top-pan balance, as for a solid of a regular shape. The volume is found by pouring water into a measuring cylinder until it is about half full. The volume of the water is read on the scale and then the pebble is carefully lowered into the water on a thin string. When the pebble is completely immersed in the water, the volume of the water is read again on the scale. The volume of the pebble is found by subtracting the first reading from the second. The density of the pebble is found by dividing the mass of the pebble by its volume.

6 The mass of a pebble was 88.4 g. The original volume of water in the measuring cylinder was 50 cm^3 and the combined volume of water and pebble was 84 cm^3. What is the density of the rock in the pebble?

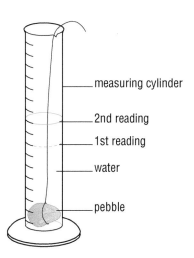

measuring cylinder

2nd reading

1st reading

water

pebble

Figure 13.3 Measuring the volume of a pebble

Measuring the density of a liquid

The density of a liquid is found in the following way.

1 A measuring cylinder is put on a balance and its mass found (A).
2 The liquid is poured into the measuring cylinder and its volume measured (V).
3 The mass of the measuring cylinder and the liquid it contains is found (B).
4 The mass of the liquid is found by subtracting A from B.
5 The density of the liquid is calculated by dividing the mass of the liquid by its volume:

$$\text{density} = \frac{B - A}{V}$$

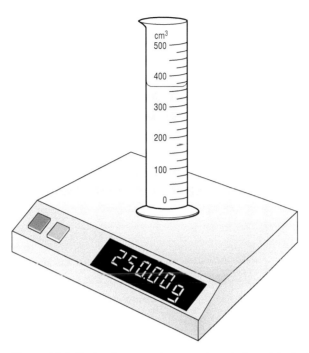

Figure 13.4 Finding the mass of a measuring cylinder and the liquid it contains

Table 13.2 shows the densities of some liquids.

Table 13.2 The densities of some liquids

Liquid	Density/kg/m^3
mercury	13550
water at 4 °C	1000
corn oil	900
turpentine	860
paraffin oil	800
methylated spirits	790

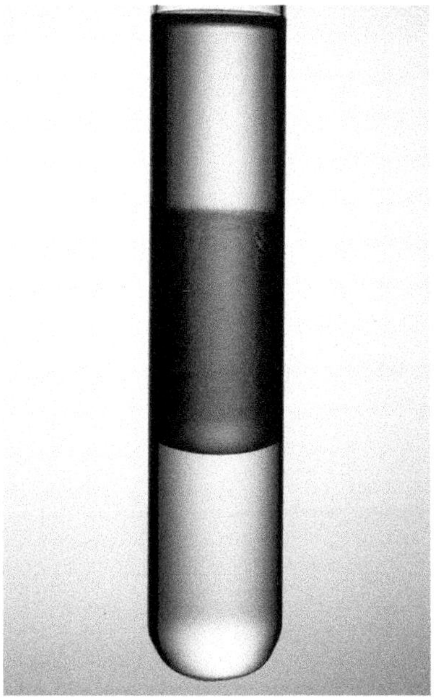

Figure 13.5 Liquids of different densities form layers when they are mixed

Floating and sinking

When a piece of wood is placed in water, the wood floats. This is due to the difference in the densities of the wood and the water. From Tables 13.1 and 13.2 you can see that wood is less dense than water. When two substances, such as a solid and a liquid or a liquid and a liquid, are put together the less dense substance floats above the denser substance.

When full-fat milk is poured into a container, such as a bottle, the cream, which contains fat and is less dense than the more watery milk, rises to the top.

7 When paraffin oil and water are poured into a container they separate and the paraffin oil forms a layer on top of the water. When water and mercury are mixed the water forms a layer on top of the mercury.

   a) What can you conclude from these two observations?

   b) What do you predict would happen if water and corn oil were mixed together?
     (Refer to Table 13.2.)

8 What do you think would happen if the following solids were placed in water?

   a) expanded polystyrene

   b) polythene

   c) perspex

Explain your answers.

(Refer to Table 13.1.)

9 What do you think would happen if the following solids were placed in mercury?

   a) steel

   b) gold

   c) lead

Explain your answers.

10 Why do you think the temperature of water is shown when the value of its density is given?

11 Most people can just about float in water (Figure 13.6). What does this tell you about the density of the human body?

12 When salt is dissolved in water the solution that is produced has a greater density than pure water. An object that floats on pure water is shown in Figure 13.7. When it is placed in salt solution do you predict that it will rise higher in the solution than it did in pure water, or sink lower?

Figure 13.6

pure water

Figure 13.7

Figure 13.8 Finding the mass of a flask of air using a top-pan balance

13 How is the process of finding the mass of a gas different from that of finding the mass of a liquid? Why is the difference necessary?

14 How can gas density be used to explain why hydrogen rises in air and carbon dioxide sinks?

Density of gases

Air is a mixture of gases. Its density can be found in the following way.

1 The mass of a round-bottomed flask with its stopper, pipe and closed clip is found by placing it on a sensitive top-pan balance. The flask is then attached to a vacuum pump and the air is removed from the flask and the clip is closed.

2 The mass of the evacuated flask, stopper, pipe and closed clip is found by placing it back on the balance. The mass of the air in the flask is found by subtracting the second reading from the first.

3 The volume of the air removed is found by opening the clip under water so that water enters to replace the vacuum. The water is then poured into a measuring cylinder to find the volume.

Table 13.3 shows the densities of some gases.

Table 13.3 The densities of some gases

Gas	Density / kg/m³
hydrogen	0.089
air	1.29
oxygen	1.43
carbon dioxide	1.98

The density of a gas changes as its temperature and pressure change. The densities of gases are compared by measuring them at the same temperature and pressure. This is called the standard temperature and pressure (STP). The standard temperature is 0 °C. The standard pressure of a gas is the pressure that will support 760 mm of mercury in a vertical tube.

When two gases meet the less dense gas rises above the denser gas.

◆ SUMMARY ◆

◆ The density of a substance is a measure of the amount of matter in a certain volume of it (*see page 170*).
◆ Common materials have a wide range of densities (*see page 171*).
◆ The density of a solid can be found by making measurements (*see page 171*).
◆ The density of a liquid can be found by making measurements (*see page 172*).
◆ The density of a gas can be found by making measurements (*see page 174*).

End of chapter questions

1 It is claimed that if you could find a lake of water large enough you could float the planet Saturn in it. The density of Saturn is 687 kg/m³. Look at Table 13.2 and decide if your agree. Explain how you came to your decision.
2 Use Table 13.3 to help you answer this question.
   a) A hydrogen balloon floats in air but would it float in the atmosphere on Mars, which has a density of 0.020 kg/m³? Explain your answer.
   b) Could an air balloon float in the Martian atmosphere? Explain your answer.
   c) The Martian atmosphere is made up mostly of one gas in Table 13.3. Which gas is it? Explain your answer.
3 If you were writing a science fiction story and wanted to have buildings floating in the atmosphere of Venus (density 65 kg/m³) which solid could you use? Look at Table 13.1 to decide.

14 Pressure

If you hold out your left hand with the palm upwards and press down on the fingertips with the fingertips of the right hand what do you feel? You may say that there is a **force** pushing down from your right hand or you may say that your fingertips are feeling a pressure. Both explanations would be correct. Now repeat with the left hand pushing on the right hand. Is there a difference? You may answer using words such as 'force' and 'pressure', but what is the connection between them? You will find out in this chapter.

Pressure on a surface

When a force (such as the push of your hand) is exerted over an area (such as the area of your fingertips) we describe the effect in terms of **pressure**. Pressure can be defined by the equation:

$$\text{pressure} = \frac{\text{force}}{\text{area}}$$

The SI unit for pressure is N/m^2 but it can also be measured in N/cm^2.

An object resting on a surface exerts pressure on the surface because of the object's weight. **Weight** is the force produced by gravity acting on a solid, a liquid or a gas, pulling the material downwards towards the centre of the Earth. The weight acts on the mass of that material. For example, the weight of a solid cube acts on that cube (Figure 14.1).

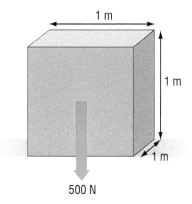

Figure 14.1 The weight acting on a cube of material

The cube pushes down on the ground (or other surface that it rests on) with a force equal to its weight. The pressure that the cube exerts on the ground is found by using the equation above. For example, if the cube has a weight of 500 N and the area of its side is 1 m², the pressure it exerts on the ground is:

$$\text{pressure} = \frac{500}{1} = 500\,\text{N/m}^2$$

If the cube has a weight of 500 N and the area of its side is 2 m², the pressure it exerts on the ground is:

$$\text{pressure} = \frac{500}{2} = 250\,\text{N/m}^2$$

An object exerts a pressure on the ground according to the area of its surface that is in contact with the ground. For example, a block with dimensions 1 m × 1 m × 2 m and a weight of 200 N will exert a pressure of 200 ÷ 1 = 200 N/m² when it is stood on one end (Figure 14.2a) but a pressure of only 200 ÷ 2 = 100 N/m² when laid on its side (Figure 14.2b).

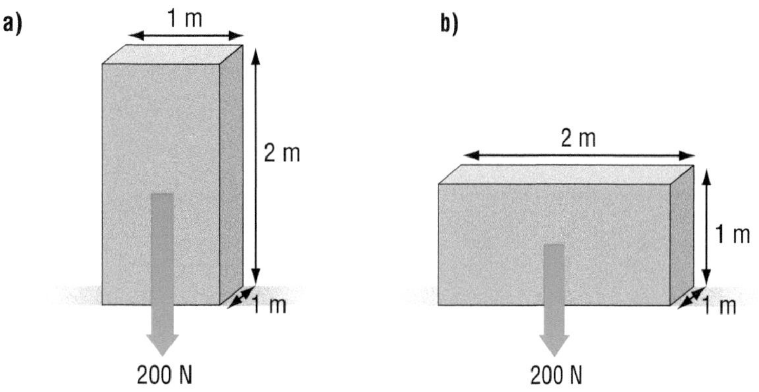

a)
1 m
2 m
1 m
200 N

b)
2 m
1 m
1 m
200 N

Figure 14.2 The weight acting on a block in two positions

1 What is the pressure exerted on the ground by a cube which has a weight of 600 N and a side area of:
a) 1 m²
b) 3 m²?

2 What is the pressure exerted on the ground by an object that has a weight of 50 N and a surface area in contact with the ground of:
a) 1 cm²
b) 10 cm²
c) 25 cm²?

3 a) What pressure does a block of weight 600 N and dimensions 1 m × 1 m × 3 m exert when it is:
i) laid on its side
ii) stood on one end?
b) Why does it exert different pressures in different positions?

Your weight acting downwards causes you to exert a force on the ground through the soles of your shoes. If you lie down, this force acts over all the areas of your body in contact with the ground. These areas together are larger than the areas of the soles of your shoes and you therefore push on the ground with less pressure when lying down than when you are standing up.

Figure 14.3 The force you exert downwards acts over a larger area when you lie down.

Reducing the pressure

When people wear skis, the force due to their weight acts over a much larger area than the soles of a pair of shoes. This reduces the pressure on the soft surface of the snow and allows the skier to slide over it without sinking.

4 Drivers in Iceland, when going out on the snow, let their tyres down until they are very soft. The tyres spread out over the surface of the snow as they drive along. Why do you think the drivers do this?

Figure 14.4 Skis stop you sinking into the snow.

Increasing the pressure

Studs

Sports boots for soccer and hockey have studs on their soles. They reduce the area of contact between your feet and the ground. When you wear a pair of these boots your downward force acts over a smaller area than the soles of your feet and you press on the ground with increased pressure. Your feet sink into the turf on the pitch and grip the surface more firmly. This makes it easier to run about without slipping while you play the game.

Pins and spikes

When you push a drawing pin into a board, the force of your thumb is spread out over the head of the pin so the low pressure does not hurt you. The same force, however, acts at the tiny area of the pin point. The high pressure at the pin point forces the pin into the board.

Sprinters use sports shoes that have spikes in their soles. The spike tips have a very small area in contact with the ground. The weight of the sprinter produces a downward force through this small area and the high pressure pushes the spikes into the hard track, so the sprinter's feet do not slip when running fast.

5 A girl wearing trainers does not sink into the lawn as she walks across it but later when she is wearing high-heeled shoes she sinks into the turf. Why does this happen?

Figure 14.5 The spikes stop the sprinter from slipping on the track. In a similar way, the studs on a soccer boot help the player to grip the turf.

Knives

As we have seen, high pressure is made by having a large force act over a small area. The edge of a sharp knife blade has a very small area but the edge of a blunt knife blade is larger. If the same force is applied to each knife, the sharp blade will exert greater pressure on the material it is cutting than the blunt knife blade and will therefore cut more easily than the blunt blade.

Figure 14.6 Knives cut well when they are sharp because of the high pressure under the blade.

Particles and pressure

Matter is made from particles. In solids the particles are held in position. In liquids the particles are free to move around each other and in gases the particles are free to move away from each other.

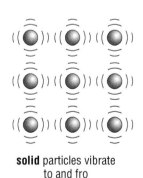

solid particles vibrate to and fro

liquid particles have some freedom and can move around each other

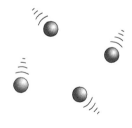

gas particles move freely and at high speed

Figure 14.7 Arrangement of particles in a solid, a liquid and a gas

Pressure in liquids

In a solid object the pressure of the particles acts through the area in contact with the ground. In a liquid the pressure of the particles acts not only on the bottom of the container but on the sides too (Figure 14.8).

a)

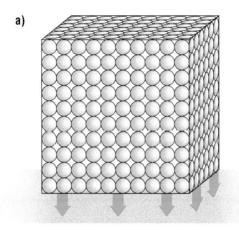

b)

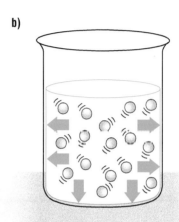

Figure 14.8 Pressure exerted by **a)** particles in a solid block and **b)** particles in a liquid

Pressure and depth of a liquid

The change in pressure with depth in a liquid can be demonstrated by setting up a can as shown in Figure 14.9. When the clips are removed from the three rubber tubes, water flows out as shown. All three jets of water leave the can horizontally but the force of gravity pulls them down. The water under the greatest pressure travels the furthest horizontally before it is pulled down. The water under the least pressure travels the shortest distance horizontally before it is pulled down.

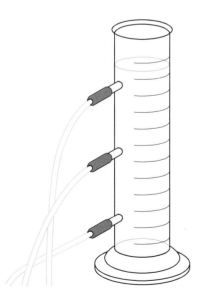

Figure 14.9 Jets of water leaving a can at different depths

6 How does the path of the jet of water at the bottom of the can in Figure 14.9 change as the water level in the can falls? Why does it change?

7 Why does a dam need a wall shaped like that in Figure 14.10?

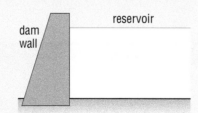

Figure 14.10 Cross-section of a dam wall

Hydraulic equipment

If pressure is applied to the surface of a liquid in a container, the liquid is not squashed. It transmits the pressure so that pressure pushes on all parts of the container with equal strength.

In hydraulic equipment a liquid is used to transmit pressure from one place to another. The pressure is applied in one place and released in another. If the area where the pressure is applied is smaller than the area where the pressure is released, the strength of the force is increased, as Figure 14.11 shows.

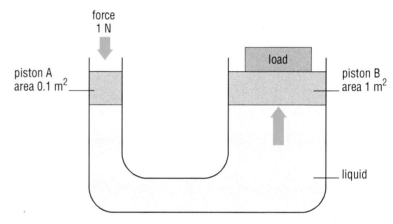

Figure 14.11 A simple hydraulic system

A car may be raised with a small force by using a hydraulic jack. When a small force is applied to a small area of the liquid in the jack, a larger force is released across a larger area and acts to raise the car.

8 Why are hydraulic systems known as 'force multipliers'?

Figure 14.12 This car has been raised into the air for repairs by a hydraulic jack.

The brake system on a car is a hydraulic mechanism. The small force exerted by the driver's foot on the brake pedal is converted into a large force acting at the brake pads. This results in a large frictional force that makes it harder for the wheels to turn and so stops the car.

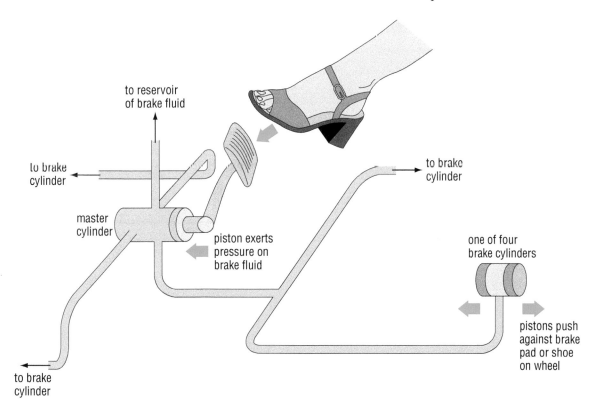

Figure 14.13 Hydraulic car brakes

Pressure in gases

Pressure of the atmosphere

The atmosphere is a mixture of gases. The molecules from which the gases are made move around but are pulled down by the force of gravity exerted on them by the Earth. The atmosphere forms a layer of gases over the surface of the Earth that is about 1000 km high. This creates a pressure of about 100 000 N/m² – equivalent to a mass of 10 tonnes on 1 m² – although this gets less as you go up through the atmosphere.

You do not feel the weight of this layer of air above you pushing down because the pressure it exerts acts in all directions, as it does in a liquid. Thus, the air around you is pushing in all directions on all parts of your body. You are not squashed because the pressure of the blood flowing through your circulatory system is strong enough to balance atmospheric pressure.

The atmosphere does not crush ordinary objects around us. For example, the pressure of the air pushing down on a tabletop is balanced by the pressure of the air underneath the table pushing upwards on the tabletop.

Ear popping

The middle part of the ear (Figure 14.14) is normally filled with air at the same pressure as the air outside the body. The air pressure can adjust because when you swallow, the Eustachian tubes in your throat open and air freely enters or leaves the middle ear. For example, if the air pressure is greater outside the body and in the mouth, when you swallow more air will enter the middle ear to raise the air pressure there.

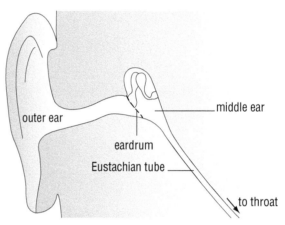

Figure 14.14 The ear and throat

9 What happens if the air pressure in the throat and outside the body is less than the air pressure in your middle ear when you swallow?

10 If you ride quickly down a hill on a bicycle your eardrums are pushed in before they pop back. Why is this?

If you travel in a car that quickly climbs a steep hill, your ears sometimes 'pop'. This is because you are rising rapidly into the atmosphere where the pressure is lower. The popping sensation is caused by the air pressure being lower in the throat and outside the body than in the middle ear. The difference in pressure causes the eardrum to push outwards. When you swallow, the air pressure in your middle ear reaches the same pressure as the air in your throat and outside, and the eardrum moves quickly back – or 'pops' – into place.

How a sucker sticks

When an arrow with a sucker on the end hits a target the arrow stays in place due to air pressure. As the elastic sucker hits the flat surface it deforms and pushes out some of the air from beneath the cup. The pressure of the remaining air in the cup is less than that of the air pressure outside the cup. The higher pressure of the air outside the cup holds the sucker in place (Figure 14.15).

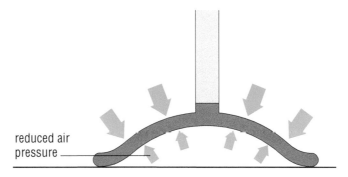

reduced air
pressure

Figure 14.15 Side view through a sucker

Crushing a can

The strength of the air pressure in the atmosphere can be demonstrated by taking the air out of a can. This can be done in two ways.

Using steam

The can has a small quantity of water poured into it and is heated from below. As the water turns to steam it rises and pushes the air out of the top of the can. If the heat source is removed and the top of the can immediately closed, the remaining steam and water vapour in the can will condense, leaving only a small quantity of air in the can. This air has a much lower pressure than the air pressure outside the can and the higher pressure crushes the can.

Using a vacuum pump

A vacuum pump can reduce the pressure in containers. If one is used to remove air from a can, the can collapses due to the greater pressure of the air on the outside (Figure 14.16).

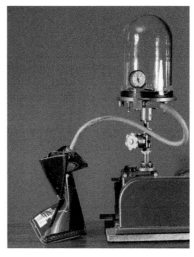

Figure 14.16 Removing air from this empty oil can has made it collapse.

A scientific showman

Otto von Guericke (1602–1686) was the mayor of Magdeburg in Germany for 35 years. He was also interested in science (page 196). He was keen to discover if a vacuum could really exist and made an air pump to test his ideas. He used his pump to draw air out of a variety of vessels. When he tried barrels he found they collapsed. He eventually found that hollow copper hemispheres joined together to make a globe were much stronger.

Aristotle had believed that if a vacuum could exist then sound would not be able to pass through it. When von Guericke put a bell in one of his vessels and removed the air he discovered that a ringing bell could not be heard.

Von Guericke greatly enjoyed demonstrating his discoveries to large numbers of people and in one demonstration had two teams of eight horses pull on the evacuated hemispheres without separating them. This spectacular demonstration helped people to realise the strength of the air pressure pushing on the hemispheres.

Figure A Otto von Guericke's demonstration with 'Magdeburg' hemispheres

1 Although von Guericke was a showman, what signs of a scientist did he also show? Explain your answer.
2 Why did the evacuated hemispheres not come apart when the teams of horses pulled on them?

3 What evidence from the past did he use to demonstrate that a vacuum had been made. Explain your answer.
4 Why did von Guericke believe he had made a vacuum?
5 How did von Guericke help people to become interested in science?

For discussion

How could an investigation you have done in your *Checkpoint Science* course be made into an informative and entertaining demonstration communicating scientific knowledge and understanding clearly?

How successful are television producers and presenters at making science programmes entertaining and informative?

Aerosols

An aerosol spray can contains a gas that is at a higher pressure than air pressure. It is held in the can by a valve in the nozzle (Figure 14.17). When the nozzle is pressed down a spring is squashed and the nozzle opening enters the inside of the can, effectively opening the valve. The higher pressure of the gas in the can pushes on the liquid in the can and it rushes up the tube and through the jet where it forms a fine spray of liquid droplets (an 'aerosol'). When the nozzle is released the spring is no longer squashed and it pushes the nozzle upwards. This removes the nozzle opening from inside the can, effectively closing the valve, and stops the flow of spray.

Aerosol cans used to contain a gas made from chlorofluorocarbons (CFCs). These chemicals are now known to damage the ozone layer. In many countries they have now been replaced with gases such as gases produced at oil refineries that do not damage the ozone layer.

11 How many uses of aerosol cans in the home can you think of?

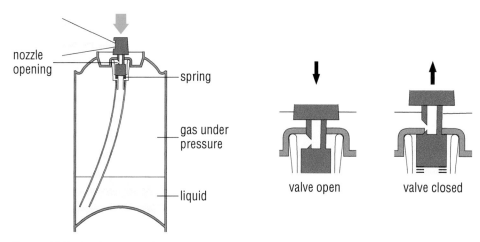

Figure 14.17 Inside an aerosol can

Hovercraft

A hovercraft uses the pressure of air to raise it from the ground. It does this by drawing air from above with powerful fans (Figure 14.18). There is a skirt around the edge of the hovercraft, which prevents the air from escaping quickly, and the air pressure beneath the hovercraft increases. The upward pressure of the air trapped beneath the hovercraft lifts the hovercraft off the ground. The fans continue to spin to replace air that is lost from the edges of the skirt.

The cushion of air beneath the hovercraft reduces friction between it and the ground. The cushion of air is also maintained when the hovercraft moves over water. The forward or backward thrust on the hovercraft is provided by propellers in the air above the hovercraft.

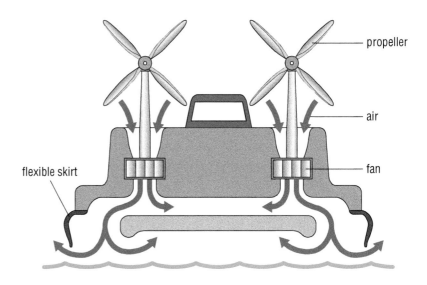

12 What are the advantages of using a hovercraft as a means of transport?

Figure 14.18 Hovercrafts work by riding on a cushion of air above the ground or water surface.

◆ SUMMARY ◆

◆ Pressure acts when a force acts over an area of surface (*see page 176*).
◆ When a solid object exerts a pressure on the surface below it, the smaller the area of contact, the greater the pressure (*see page 177*).
◆ Pressure in a liquid acts in all directions and increases with the depth of the liquid (*see page 181*).
◆ In hydraulic systems pressure is transmitted through a liquid (*see page 182*).
◆ The atmosphere exerts a pressure (*see page 183*).
◆ Use is made of the pressure of air in various devices such as suckers, pumps and hovercraft. Use is made of the pressure of other gases in aerosols (*see pages 184–188*).

End of chapter question

How could you explain the following, using the model of air made up of particles which move freely?
a) How air pushes on a surface.
b) Why the pressure in an inflated tyre is higher than the air pressure outside.
c) Why a sucker stays in place on a flat surface.

15 Turning on a pivot

◆ The turning effect of forces
◆ Types of levers
◆ Moments
◆ The principle of moments

At the beginning of Chapter 7, we saw how Ancient Greek philosophers argued about their ideas from what they saw of the world around them. There was one philosopher who was very different. He was called Archimedes (287–212 BCE) and where other philosophers argued he set about finding out about the world by making measurements. The best example of his measurements in science is in his study of the **lever**. He did not invent the lever – it had been known from the earliest times as a device that can help lift an object. He was the first person to study it scientifically and his work inspired many scientists who came after him – most notably Galileo whose work you have studied in the introductory chapter and Chapter 17 of *Student's Book 1*.

In this chapter, we look at a law Archimedes discovered and examples of the three types of levers.

Figure 15.1 Archimedes was so excited about his work on the lever he stated that if he had a long enough lever he could move the Earth.

Figure 15.2 Tightening a nut

The turning effect of forces

A force can be used to turn an object in a circular path. For example, when you push down on a bicycle pedal the cog wheel attached to the crankshaft turns round.

A nut holding the hub of a bicycle wheel to the frame is turned by attaching a spanner to it and exerting a force on the other end of the spanner in the direction shown in Figure 15.2.

A device that changes the direction in which a force acts is called a lever. It is composed of two arms and a **fulcrum** or **pivot**. The lever also acts as a force multiplier. This means that a small force applied to one arm of the lever can cause a large force to be exerted by the other arm of the lever. For example, a crowbar is a simple lever. It is used to raise heavy objects. One end of the crowbar is put under a heavy object and the crowbar is rested on the fulcrum (Figure 15.3). When a downward force is applied to the long arm of the crowbar, an upward force is exerted on the heavy object. A small force acting downwards at a large distance from the fulcrum on one side produces a large force acting upwards a short distance from the fulcrum on the other arm.

The force applied to the lever to do the work is called the **effort**. It opposes the force that is resisting the movement, called the **load** (Figure 15.3).

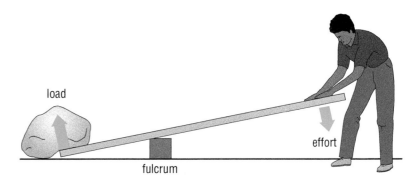

Figure 15.3 Using a lever: a simple crowbar

Types of levers

There are three types of levers, shown in Figure 15.4. They are based on the relationship between the effort, the load and the points of action relative to the fulcrum.

1 What type of lever is a crowbar?
2 In the human arm the elbow is the fulcrum, the effort is produced by the biceps and the load acts downwards through the hand. What class of lever is this part of the human arm?
3 A bottle opener has its fulcrum at one end of the lever and the effort is applied at the other end. What class of lever is it?

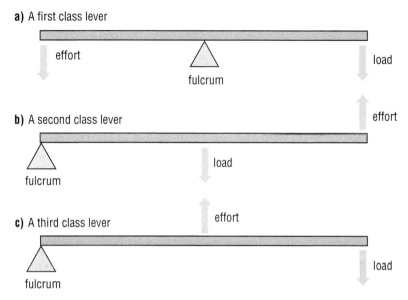

a) A first class lever

effort
fulcrum
load

b) A second class lever

effort
fulcrum
load

c) A third class lever

effort
fulcrum
load

Figure 15.4 Three classes of levers

Moments

The turning effect produced by a force around a fulcrum is called the **moment** of the force. This is best understood by referring to the first class lever in Figure 15.4a as you read the following text.

The direction of the moment is usually specified as clockwise (the load in Figure 15.4) or anticlockwise (the effort in Figure 15.4) about the fulcrum. The size of the moment is found by multiplying the size of the force by the distance between the point at which the force acts and the fulcrum (in Figure 15.4a these are the distances from the ends to the fulcrum). The moment of a force can be shown as an equation:

moment of force = force × distance from the fulcrum

The moment is measured in newton metres (Nm). The moment of the force applied to one arm of a lever is equal to the moment of the force exerted by the other arm. For example, a 100 N force applied downwards 2 m from the fulcrum produces a 200 N force upwards 1 m from the fulcrum on the other arm.

4 What is the moment of a 100 N force acting on a crowbar:
  a) 2 m from the fulcrum
  b) 3 m from the fulcrum
  c) 0.5 m from the fulcrum?

5 A 100 N force acting on a lever 2 m from the fulcrum balances an object 0.5 m from the fulcrum on the other arm. What is the weight of the object (in newtons)? What is its mass (in kg)?

6 Where will the strongest force be exerted by scissor blades to cut through a piece of material? Explain your answer.

7 Why can a lever be described as a force multiplier?

In the case of a seesaw, which is another simple type of lever, the moment of the weight on one arm must equal the moment of the weight on the other arm for the seesaw to balance. From this observation, Archimedes constructed his law of moments, which is sometimes called the law of the lever or the principle of moments. This law or principle states that when a body is in equilibrium (or balance) the sum of the clockwise moments about any point (such as the fulcrum) equals the sum of the anticlockwise moments about that point.

A pair of pliers (Figure 15.5) is made from two levers. When they are used to grip something a small force applied to the long handles produces a large force at the short jaws.

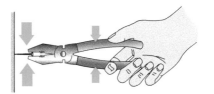

Figure 15.5 Using pliers and scissors

◆ SUMMARY ◆

◆ A force can produce a turning effect (*see page 190*).
◆ A lever is a device that changes the direction in which a force acts (*see page 190*).
◆ There are three types or classes of levers (*see page 191*).
◆ The turning effect produced by a force around a fulcrum is called a moment (*see page 191*).

End of chapter question

Is it possible to balance a mass of weight 5 N with a mass of weight 15 N on a model seesaw with 10 cm arms? Explain your answer.

Suggest where you might place each mass to get an exact balance.

16 Electrostatics

◆ The atom and electric charge
◆ Charging materials
◆ Insulators and conductors
◆ Induced charges
◆ Sparks and flashes
◆ The van de Graaff generator
◆ Digital sensors

You may have seen a balloon stuck to the wall or ceiling (Figure 16.1), or received a slight shock when touching a car door, or heard a crackle when taking off your jumper. To understand why these things happen, you have to think about the structure of the atom and the electric charges on the particles in it.

The atom and electric charge

Figure 16.1 These balloons are held in place by electrostatic forces.

An atom has a central nucleus surrounded by electrons and each electron carries a negative electric charge. In the nucleus are particles called protons and each proton carries a positive electric charge. Usually the number of positive charges carried by the protons is balanced by the number of negative charges carried by the electrons. For example, if an atom has six protons in its nucleus it has six electrons orbiting the nucleus. When the positive charges on the protons are balanced by the negative charges on the electrons in this way the atom is described as being neutral.

As you can see in Figure 16.2, there are also particles called neutrons in the nucleus. They have no electric charge.

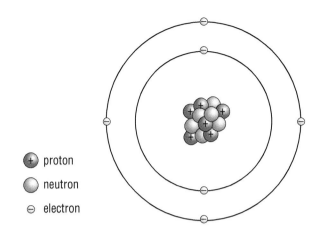

Figure 16.2 The structure of an atom

Charging materials

In the party trick with the balloons (Figure 16.1), each balloon must be rubbed on clothing such as a woollen sleeve, before it will stick to the wall. When some dry materials are rubbed in this way they gain electrons from the atoms in the material they are being rubbed against (Figure 16.3). Other materials lose electrons to the material they are being rubbed against. It depends upon the particular pair of materials involved. When a material that is an electrical insulator gains or loses electrons in this way, it is left with excess charge and the charge stays in place when the materials are separated. The material has been charged with static electricity.

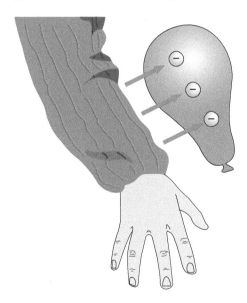

Figure 16.3 Electrons are transferred from the wool to the balloon and stay there.

1 When a piece of polythene is rubbed with a dry woollen cloth, electrons move from the cloth to the polythene. Which material becomes:
   a) positively charged
   b) negatively charged?

2 When a piece of perspex is rubbed with a dry woollen cloth, electrons move to the cloth. Which material becomes:
   a) positively charged
   b) negatively charged?

3 If a charged piece of polythene is set up as shown in Figure 16.5 and a charged piece of perspex is brought close to it, will the polythene swing towards the perspex or away from it? Explain your answer.

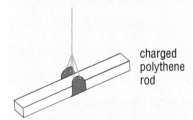

charged polythene rod

Figure 16.5

4 If a charged strip of polythene is set up as shown in Figure 16.5 and a charged polythene rod is brought close to it, will the polythene strip swing towards the polythene rod or away from it? Explain your answer.

5 **a)** When long dry hair is brushed the strands often move away from each other. Why do you think this happens?
   b) The strands of hair also get attracted to the brush. Why do you think this happens?

A material that gains electrons when it is rubbed has more negative charges than positive charges and so is said to be negatively charged. A material that loses electrons when it is rubbed has fewer negative charges than positive charges and so is said to be positively charged. Protons are never transferred in this charging process since they are effectively locked in place in the nuclei of the atoms of the material.

When the balloon is rubbed on a sleeve it receives electrons from the material in the sleeve and its surface becomes negatively charged. If two balloons, suspended on nylon threads, are charged and placed close to each other they move apart. The negative charges on the balloons repel each other. If a positively charged material is placed close to a negatively charged balloon hanging on a nylon thread the balloon moves towards the material because the different charges on the two materials attract each other (Figure 16.4).

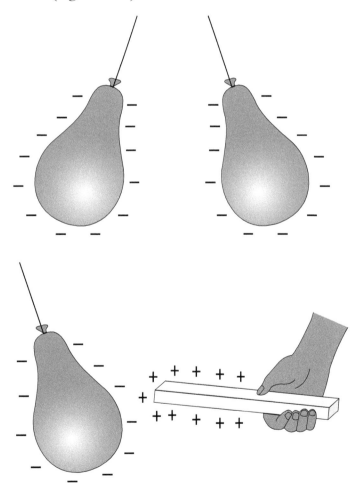

Figure 16.4 Similar charges repel (top) and different charges attract (bottom).

Early studies of electricity

For millions of years, there have been certain kinds of trees producing sap that turns to a clear yellow fossilised substance called amber. In Ancient Greece, amber was used in items of jewellery. Thales (624–546 BCE), the earliest Greek 'scientific' philosopher, noticed that if amber was rubbed it developed the power to pick up small objects like dust, straw and feathers. The Greek word for amber is elektron so, much later, its attractive power became known as electricity.

William Gilbert (1544–1603), an English scientist, discovered that a few other materials, such as certain gemstones and rock crystal, could also attract small objects when they were rubbed. He called these materials 'electrics'.

A German scientist called Otto von Guericke (1602–1686) used an 'electric' called sulfur to make a machine that could generate sparks. He made the sulfur into a ball and attached it to an axle that could be turned quickly by a crank handle. As the ball span, it was rubbed and built up a charge of static electricity, which produced sparks. Electric machines became popular as a form of amusement and entertainment. Some people made their living by travelling through European countries, demonstrating their machines.

Figure A An electric machine built in about 1762

Stephen Gray (1696–1736), another English scientist, investigated electricity by rubbing a glass tube that had corks in the ends, and discovered that the corks became charged with static electricity even though they had not been touched. He had discovered that electricity behaves as if it can flow like a liquid.

1 Where did the word 'electricity' come from?
2 Why do you think electric machines amused people?

3 How was the evidence provided by Thales' observation used by Gilbert to extend the knowledge of electricity?

Charles Du Fay (1698–1739), a French scientist, repeated Gray's experiments and extended them by comparing the way in which the objects were charged with electricity. He discovered that if he charged a cork ball using a glass rod that had been rubbed, it was attracted to a cork ball that had been charged using sealing wax that had been rubbed. He also discovered that two cork balls that had both been charged by either the rubbed glass or by the rubbed sealing wax repelled each other. From his investigations he believed that electricity was made from two different fluids. They were called 'vitreous electricity' (from rubbing glass) and 'resinous electricity' (from rubbing sealing wax).

Benjamin Franklin (1706–1790), an American statesman and scientist, refined Du Fay's idea of two electrical fluids by suggesting that when substances were charged they received either too much fluid and became positively charged or they had some fluid taken away and became negatively charged.

4 How was the evidence provided by Gray's work used by Du Fay to secure and extend the knowledge of electricity?

5 What creative thoughts did Franklin apply to Du Fay's work to refine ideas about electricity?

6 a) How was Franklin's idea of electric charge similar to the ideas that we use today?

b) How was his idea of electric charge different to those we use today?

Insulators and conductors

A material that can become charged with static electricity is called an **insulator**. If electrons are added to the material they stay in place and the insulator is negatively charged. If electrons are removed from the material more electrons do not flow into the material and it remains positively charged.

A metal cannot be charged with electricity by rubbing, in the way that an insulating material can, because electrons flow easily through metals. A material through which electrons can flow is called a **conductor**. The human body is a very good conductor of electricity.

Induced charges

If a material has an electric charge, it can make or 'induce' an electric charge on the surface of a material close by without touching the material. For example, if a piece of plastic, such as a pen, is rubbed and held above a tiny piece of paper, the positive charge on the plastic draws electrons to the surface of the paper nearest the plastic. This makes the uppermost surface of the paper negatively charged. When the pen is brought very close to the paper the force of attraction between the two surfaces is strong enough to overcome the weight of the paper and the paper springs up to the surface of the pen.

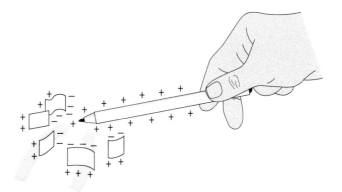

Figure 16.6 The charged pen induces charges on the surfaces of the paper scraps.

The underside of the paper is left with a positive charge but since this is further away from the pen the force of repulsion it experiences is weaker than the attractive force and the paper is held.

In a similar way a charged balloon induces an opposite charge on the surface of a wall it is brought close to. When the balloon touches the wall, the force of attraction between the two surfaces is greater than the weight of the balloon and the air inside it, so the balloon sticks to the wall (Figure 16.1). This way of charging a material without touching it is called charging by induction.

6 When a negative charge is induced on one surface of a piece of paper, what is induced on the other surface of the paper? Why does this happen?

7 Why does a rubbed balloon stick to the wall?

Sparks and flashes

Air is a poor conductor of electricity but if the size of the charge on two oppositely charged surfaces is very large the air between them may conduct electricity as a spark or a flash, like a flash of lightning. This happens when the molecules in the air are split. They form negatively charged electrons and positively charged ions. The electrons move towards the positively charged surface and ions move towards the negative surface (Figure 16.7). As the electrons move they collide with other molecules in the air and split them. The ions and electrons from these molecules also move towards the charged surfaces and split more molecules as they go. This process occurs very quickly and produces a spark.

When the charged particles in the air meet the charged surfaces, the positive and negative charges cancel each other out and the surfaces lose their charge. They are said to have been discharged.

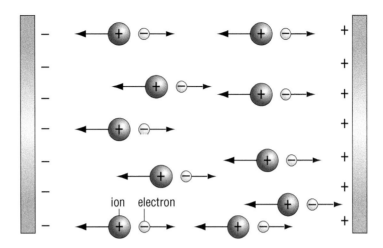

Figure 16.7 The strong electric field between the charged plates ionises the air between them.

ion   electron

Preventing an explosion

When an aircraft flies through the air, its surfaces are rubbed by air particles and become charged with static electricity. If the aircraft was equipped with non-conducting tyres, such as those used on most vehicles, the charge would remain on the aircraft when it landed. This charge could cause a spark during refuelling. The heat from the spark would be sufficient to cause the fuel vapour to combust, which would result in a devastating explosion. This danger is prevented by equipping the aircraft with tyres that conduct electricity. When the aircraft lands the charge it possesses passes to the ground.

Figure 16.8 Sparks must be prevented when an aircraft is being refuelled.

Lightning

When a storm cloud develops, strong winds move upwards through the cloud and rub against large raindrops and hail stones. This rubbing causes the development of charged particles in the cloud. Positively charged particles collect at the top of the storm cloud and negatively charged particles collect at the base. The size of the different charges in each part of the cloud may become so large that lightning, called sheet lightning, is produced between them.

The negative charge at the base of the storm cloud induces a positive charge on the ground below. If the charges become large enough a flash of lightning, called forked lightning, occurs between them (page 169).

The van de Graaff generator

In 1931, Robert van de Graaff invented a machine that produced a huge charge of static electricity. The machine is called the van de Graaff generator.

In the generator is a rubber belt that runs over two rollers. One roller is made of perspex and the other is made of polythene. The belt is driven by an electric motor. When the perspex roller is placed at the base of the generator, the belt running over it becomes negatively charged (Figure 16.9). The charged part of the belt rises to the polythene roller at the top of the generator where there is a device that transfers the negative charge to the hollow metal dome. The belt moves over the roller in the dome and back to the roller at the base, where it becomes negatively charged again.

When the rollers are reversed and the polythene roller is placed at the base of the generator, the belt becomes positively charged and the positive charge is transferred to the dome.

8 a) Explain what you see in Figure 16.10.
   b) Why must the student be standing on a sheet of insulating material?

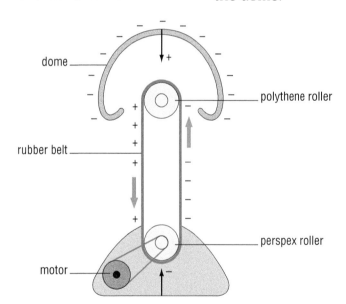

Figure 16.9 Inside a van de Graaff generator

Figure 16.10 Demonstrating a large electrostatic charge with a van de Graaff generator

Very high charges can be stored in the dome and released during investigations. In the past, huge van de Graaff generators were used as particle accelerators to investigate atomic structure but they were replaced by other devices. Today they are used in schools and colleges to generate quite large electrostatic charges for demonstration lessons.

Digital sensors

As scientists studied electrostatics they discovered a way of storing charge and releasing it when required. The charge-storing device is called a **capacitor** and a simple example is shown in Figure 16.11. The charge is stored on the metal plates and the dielectric is a material that stops the charge from crossing between them. The charge is released when a conductor is switched into a circuit between the plates and a current flows.

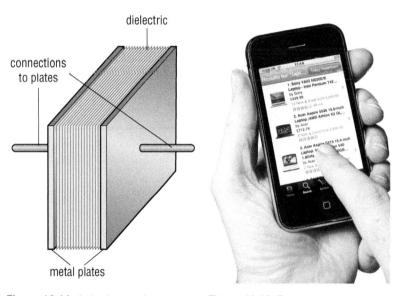

Figure 16.11 A simple capacitor

Figure 16.12 Touch screen technology would not have been possible without the development of the capacitor.

Since the development of the simple capacitor using metal plates tiny, electronic devices have been invented to act as switches, conductors and capacitors. These are now used in all kinds of electronic apparatus, from sensors used in laboratories to detect changes in light, pH and movement, to the touch screens on mobile phones, satellite navigation devices and other computer displays, where your fingertip and the screen act like the plates of a capacitor and the tiny difference in charge between them is used by the computer in the appliance to trigger a response.

◆ SUMMARY ◆

◆ An atom has two kinds of electrically charged particles. They are protons and electrons (*see page 193*).

◆ Static electricity may be generated by rubbing insulators (*see page 194*).

◆ Charged materials can have either a positive or a negative charge (*see page 195*).

◆ Materials that hold charges of static electricity are called insulators. Materials that allow electricity to pass through them are called conductors (*see page 197*).

◆ A material can become charged by induction (*see page 197*).

◆ Static electricity can be discharged through the air as a spark or a flash (*see page 198*).

◆ The van de Graaff generator can produce large charges and can be used to study the effects created by static electricity (*see page 200*).

◆ Electrostatic devices are used in digital sensors (*see page 201*).

End of chapter question

How could you use your knowledge of the structure of the atom to explain how a plastic pen that has been rubbed can pick up tiny pieces of paper?

◆ Simple circuits
◆ Other circuit components
◆ Amperes
◆ Measuring current
◆ Voltage
◆ Measuring voltage
◆ Resistance

For discussion

Name the electrical things you use during the course of a day from morning to evening. How would your life change without them?

Electricity is so much part of our lives that we probably never think about it. In this chapter, though, we will look at electricity closely to make sure we understand what it is and how it can help us do so much.

Simple circuits

If you set up this equipment and close the switch, the lamp comes on.

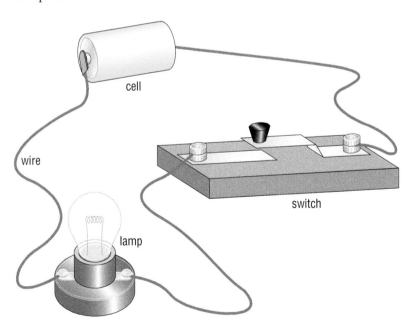

Figure 17.1 A simple circuit

1 Describe the path of an electron round the circuit in Figure 17.1 (page 203) when the switch is pressed down.

2 **a)** How does the wire in the filament behave differently to other wires in the circuit when the current flows?

   b) What property of the wire accounts for this difference?

The wires of the circuit are composed of atoms that are held tightly together but around them are many electrons that are free to move. The metal filament in the lamp and the metal parts of the switch also have free electrons. When the switch is closed, the wires on either side of the switch are linked by metal contacts and a path is made along which the electrons can flow. When you open the switch, the lamp goes out. The path is broken and the electrons cannot flow.

The energy to move the electrons comes from the cell. The chemical reactions that take place in the cell make the electrons leave the cell at the negative terminal when the circuit is completed. They push their way into the wire and move the other electrons along, creating a flow or **current** of electricity. At the positive terminal electrons are drawn back inside the cell.

The wire in the lamp filament is more resistant (page 206) to the flow of electrons than the other wires in the circuit. As the current moves through the filament some of its electrical energy is transferred to heat energy and light energy.

In time the chemicals that take part in the reaction inside the cell are used up. They can no longer release energy to make the electrons move and the current stops. The number of electrons in the circuit does not change – it is the chemical energy released by the cell that changes.

When Benjamin Franklin (page 197) described substances as being either positively or negatively charged, he thought that electricity ('electrical fluid') flowed from a positively charged substance to a negatively charged one. His idea was taken up by other scientists until it was discovered that it was the flow of negatively charged electrons that produced a current. Franklin's idea is still used today, however – it is known as the conventional current direction.

When circuits are drawn, symbols are used for the parts or components. The use of symbols instead of drawings makes diagrams of circuits quicker to make and the connections between the components are easier to see. The symbols have been standardised, like the SI units, and are recognised by scientists throughout the world. The circuit in Figure 17.1 is shown as a circuit diagram using symbols in Figure 17.2. The components for the circuit are the wires, cell, lamp and switch.

3 In Figure 17.1 (page 203), the base of the cell (on the right) is the negative terminal and the cap (on the left) is the positive terminal. How can you distinguish between the positive and negative terminals in a cell in the circuit diagram in Figure 17.2?

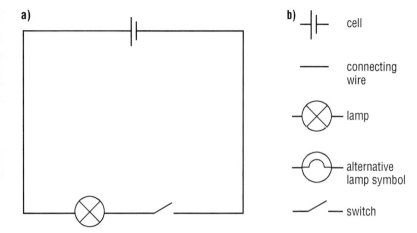

Figure 17.2 A circuit diagram and the symbols used

In everyday life, cells are almost always called batteries but this is scientifically incorrect. In science, a battery is made of two or more cells joined together. The symbols for batteries of two cells, three cells and more cells are shown in Figure 17.3.

a) b) c)

Figure 17.3 The circuits symbols for two cells, three cells and any number of cells

Other circuit components are also represented by unique symbols, as you will see in this chapter – for example, resistor (page 206) and variable resistor (page 207), buzzer (page 207), ammeter (page 209), voltmeter (page 213), light-dependent resistor (page 214), diode and light-emitting diode (page 215).

The lining up of cells next to each other in a row, end to end, as shown in Figure 17.3, is described as arranging them in series. Other electrical components can also be arranged in series, as you will see later.

A solid can be tested to find out if it conducts electricity by using a circuit like the one shown in Figure 17.4. The solid to be tested is secured between the pair of crocodile clips and the switch is closed. The lamp lights up if the solid conducts electricity. By using the circuit, metals and the non-metal carbon, in the form of graphite, are found to be conductors of electricity. Other non-metals such as sulfur and solid compounds such as sodium chloride do not conduct electricity. They are insulators. Materials such as wood and plastic are also insulators.

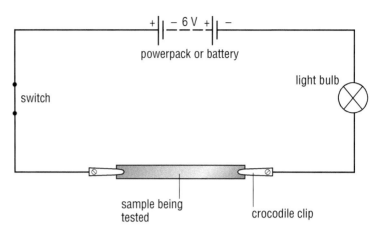

Figure 17.4 A circuit for testing conduction of solid materials

Resistance

The material through which a current flows offers some **resistance** to the moving electrons. A material with a high resistance only allows a small current to pass through it when a certain voltage is applied. A material with a low resistance allows a larger current to pass through it for the same applied voltage (see page 212).

Other circuit components

Resistors

4 How does the length of a high resistance wire affect the flow of current through the circuit?

If a short piece of wire with a high resistance is included in a simple circuit with a cell, a switch and a lamp, the lamp will shine less brightly than before. If a longer length of high-resistance wire is included in the circuit the light will shine even more dimly.

A component that is designed to introduce a particular resistance into a circuit is called a **resistor**.

Figure 17.5 Four resistors and the symbol for a resistor

5 a) In Figure 17.6, which way should the contact be moved to increase the resistance in the part of the wire included in the circuit between A and B?

b) Which way should the contact be moved to decrease the resistance in the part of the wire included in the circuit between A and B?

6 Figure 17.8 shows a variable resistor in a dimmer switch. How would you turn the switch to make the lights:

a) brighter

b) dimmer?

Explain your answer.

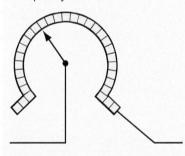

Figure 17.8

7 Draw a circuit diagram for circuits that have:

a) two cells, a variable resistor, two lamps and a switch

b) three cells, a lamp, a buzzer and a switch

c) two cells, a variable resistor, a buzzer and a switch.

A variable resistor can be made in which a contact moves along the surface of a resistance wire and brings different lengths of the wire into the circuit. This device is sometimes called a rheostat. In order to make it more compact, the length of the wire is wound in a coil and the contact is made to move freely across the top of the coil.

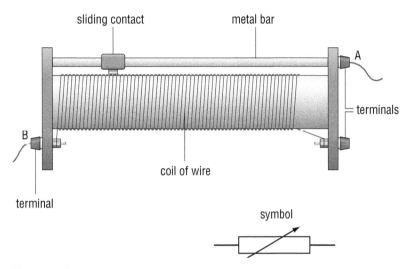

Figure 17.6 A variable resistor and its symbol

In Figure 17.6 the current passes through terminal A, along the bar, through the sliding contact and coil of wire to terminal B. When the contact is placed on the far left the current passes through only a few coils of the wire. As the contact is moved to the right the current flows through more of the wire and encounters greater resistance. When the contact is moved from the right to the left the current flows through fewer coils of the wire and encounters less resistance.

Buzzers

A buzzer is an electrical device in which one part vibrates strongly when an electric current passes through it. The vibrations produce the sound.

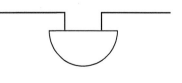

Figure 17.7 A buzzer and its symbol

Fuses and circuit breakers

When a current flows through a circuit, some energy is lost as heat (thermal) energy. If the size of the current increases, the amount of heat (thermal) energy released also increases. A **fuse** is a device containing a wire that melts when the current flowing through it reaches a certain value. When the wire melts, it breaks and so also breaks the circuit and stops the current flowing.

Electrical appliances are designed to work when a current of a certain size flows through them. If the current is too large, the appliance may be damaged. An unusually large current can occur in a household circuit in two ways. It may occur when the insulation in a cable is worn and the wires in the cable touch each other. This causes a short circuit. It may also occur if too many appliances are plugged into one socket. Fuses are used to stop the flow of a current when it becomes too large for the circuit. They may be present in plugs (Figure 17.9) and/or in the appliances themselves. In the past each circuit in a home was protected by a fuse in a consumer unit or 'fuse box'. Today, circuit breakers are used instead of fuses in consumer units. A circuit breaker is a switch that is sensitive to the size of the current flowing through it. If the current is too large, the switch opens and breaks the circuit. The switch can be closed and the circuit used again once the cause of the problem has been identified and corrected.

8 How could you prevent a fire in the home being caused by an electrical fault?

9 Why should you always switch off a circuit before replacing a fuse?

a)

b)

c)

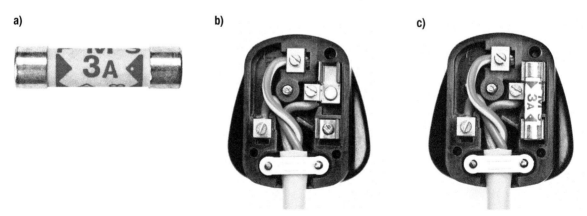

Figure 17.9 a) A three-amp fuse **b)** the inside of a plug with no fuse **c)** the inside of a plug with a fuse in place.

Amperes

The rate at which electrons flow through a wire is measured in units called amperes. This word is usually shortened to amps and the symbol for it is A. One amp is equal to the flow of 6 million, million, million electrons passing any given point in the wire in a second!

Measuring current

The current flow in a circuit is measured using an instrument called an **ammeter**. This is a device that has a coil of wire set between the north and south poles of a magnet. The coil has a pointer attached to it and it turns when a current passes through it. The amount by which the coil turns depends on the size of the current and is shown by the movement of the pointer across the scale.

When an ammeter is used it is connected into a circuit with its positive (red) terminal connected to a wire that leads towards the positive terminal of the cell, battery or power pack. It is always connected in series with the component through which the current flow is to be measured (Figure 17.10). Ammeters usually have a very low resistance so that the current passes through them without affecting the rest of the circuit.

10 To think about current and electron flow, try these simple calculations. How many electrons are flowing per second past a point in a circuit in which there is a current of:
  a) 0.5A
  b) 5A
  c) 30A?
11 Towards which terminal of the power supply should the negative (black) terminal of an ammeter be connected?

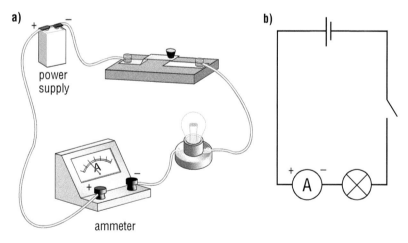

Figure 17.10 a) An ammeter connected in a circuit and **b)** the circuit diagram showing its symbol

When an ammeter is to be used to measure the current flowing through a series circuit such as that shown in Figure 17.11a, the ammeter is placed at a position such as A or B. When an ammeter is to be used to measure the current flowing through a parallel circuit (page 211) such as that shown in Figure 17.11b, the ammeter should be placed at A, B and C in turn.

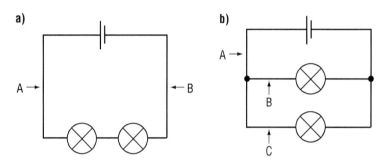

Figure 17.11 Measuring current in **a)** series and **b)** parallel circuits

The current in series circuits

In a **series circuit**, as shown in Figure 17.12, the ammeter will record the same amount of current in each position A, B, C and D.

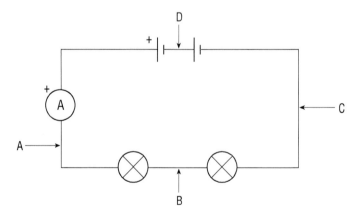

Figure 17.12 Measuring current in a series circuit using an ammeter

12 A wire carrying a current of electricity can be described as being similar to a stream carrying a current of water. In what ways are the wire and the stream similar?

13 Predict the brightness of the lamps in the circuits in Figure 17.14 compared with that of a single lamp in a circuit with one cell. Use one of the following descriptions in each case: very dim, dimmer, the same, brighter, very bright. (All the lamps are identical and all the cells have the same voltage.)

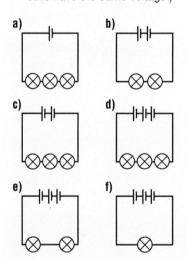

Figure 17.14

14 Compare the circuit in Figure 17.15 with the one in Figure 17.14b. Do you think the lamps will glow with the same brightness? Explain your answer.

Figure 17.15

Lamps and current size

The wires connecting the components in a circuit have a low resistance while the wires in the filaments of lamps have a high resistance. When lamps are connected in series, their resistances combine – they add. They therefore offer a greater resistance to the current than each lamp would separately.

The size of the current flowing though a circuit can be estimated by looking at the brightness of the lamps in the circuit. A lamp shines with normal brightness when it is connected to one cell as shown in Figure 17.13a. The lamp shines more brightly than normal when it is in a circuit with two cells (Figure 17.13b) and shines less brightly when it is in a circuit with one cell and another lamp as shown in Figure 17.13c.

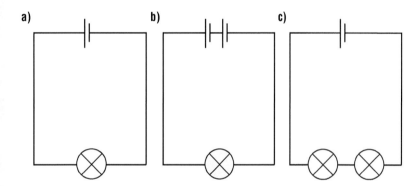

Figure 17.13 Three arrangements of cells and lamps in series

The current in parallel circuits

Lamps can be arranged in a circuit 'side by side' rather than end to end. This kind of circuit is called a **parallel circuit** (Figure 17.16). The resistances of the lamps do not combine to oppose the flow of current in the same way as they do in a series circuit. Each lamp receives the same flow of electrons as it would if it were on its own in a circuit with the cell.

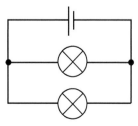

Figure 17.16 Two lamps in parallel

15 How do you think the brightness of two lamps arranged in parallel compares with the brightness of two lamps arranged in series (both arrangements having one cell)?

16 a) If current flows through two lamps arranged in series, and the filament of one lamp breaks, what happens to the other lamp? Explain your answer.

b) If the two lamps are arranged in parallel, and the filament of one lamp breaks, what happens to the other lamp? Explain your answer.

In a parallel circuit like the one shown in Figure 17.17 the ammeter reading varies in the following way. At points A and D the readings are the same. At points B and C, if the lamps are identical the readings will be the same and exactly half the readings at A and D. This means that if the readings at A and D are 4.0A the readings at B and C will be 2.0A. The total of the readings of the ammeters at B and C is always the same as the reading at A and D even if the bulbs are not identical. For example, if the reading at A and D is 4.0A and the reading at B is 3.0A then the reading at C will be 1.0A.

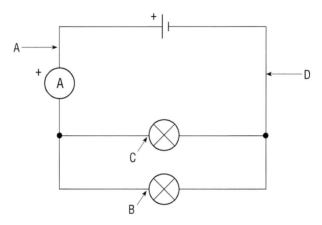

Figure 17.17 Using an ammeter to measure the current at different points in a parallel circuit

Voltage

The ability of the cell to drive a current is measured by its **voltage**. This is indicated by a figure on the side of the cell with the letter V after it. The volt, symbol V, is the unit used to measure the difference in electrostatic potential energy, usually just referred to as potential difference, between two points. The voltage written on the side of the cell refers to the difference in potential between its positive and negative terminals. It is a measure of the electrical energy that the cell can give to the electrons in a circuit.

When cells are arranged in series with the positive terminal of one cell connected to the negative terminal of the next cell, the current-driving ability of the combined battery of cells can be calculated by adding their voltages. For example, two 1.5V cells in series produce a voltage of 3V. The two cells together give the electrons in the circuit twice as much electrical energy as each one would provide separately.

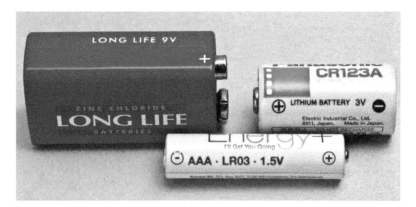

Figure 17.18 The voltage is clearly displayed on the packaging of cells and batteries.

Measuring voltage

The voltage or potential difference between two points in a circuit is measured using a **voltmeter**.

17 Compare how an ammeter and a voltmeter are connected into a circuit by looking at Figure 17.10 on page 209 and Figure 17.19 on this page.

18 a) In a series circuit, the voltage across the components can vary. For example, in a circuit in which three lamps are joined in series the voltages were 0.4V, 0.5V and 0.6V. The cell providing the electrical power to the circuit has a voltage of 1.5V. From this, what can you say about the total voltage of the components in the circuit and the voltage of the power source?

b) In a parallel circuit, the voltage across all the components in a circuit is the same. If three lamps were arranged in a parallel circuit powered by a 1.5V cell, what would be the voltage across each lamp?

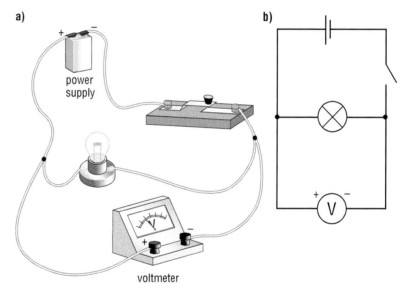

Figure 17.19 a) A voltmeter connected in a circuit and **b)** the circuit diagram showing its symbol

The voltmeter is connected into a circuit with its positive (red) terminal connected to a wire that leads towards the positive terminal of the cell, battery or power pack. The negative (black) terminal must be connected to a wire that leads towards the negative terminal of the source of the current. However, unlike the connection of an ammeter, the wires are attached to either side of the part of the circuit being tested – it is arranged in parallel with this part of the circuit. Voltmeters generally have a very high resistance, so when connected in parallel they take little current and do not affect the rest of the circuit.

Semiconductors and electrical components

Ferdinand Braun (1850–1918), a German physicist, performed many experiments on crystals. He attached pairs of metal contacts to a crystal and attached them to a voltage generator to see what would happen. He found that in many crystals a current between the contacts was not produced which meant that the crystal was an insulator. He also found that when he tested a few crystals a current was produced between the contacts which meant that the crystal was a conductor.

Braun went further and used very tiny contacts made from needles on crystals that conducted electricity. He found that they conducted electricity as before but when he reversed the voltage, the current of electricity would not flow in the opposite direction. The crystal had become an insulator. Crystals that behave like this became known as semiconductors. They were used in the first radios, which were sometimes called crystal sets.

An American physicist called William Shockley (1910–1989) knew of Braun's work and worked with other scientists in a team to take Braun's investigations further. They discovered that if semiconductor crystals were given an impurity of certain other elements they could be developed to make very precise devices for controlling the flow of electricity. Today these semiconductors are used in a wide range of electronic components as the following examples show.

Light-dependent resistors

A light-dependent resistor (LDR) is made from two pieces of metal, joined together by a semiconductor. A semiconductor is a material with just a few electrons that can move freely.

When the LDR receives light energy, more electrons are released in the semiconductor to move freely through it, and the resistance of the LDR becomes lower. When the amount of light shining on it is reduced, fewer electrons can flow and the resistance increases.

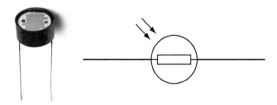

Figure A A light-dependent resistor (LDR) and its symbol

1 After testing a few crystals and finding they did not conduct electricity, what do you think Braun may have predicted for every other experiment he was about to make?

2 What statement of considering evidence and approach did Braun carry out when he used tiny contacts instead of the ordinary ones he had used before?

3 What evidence did Shockley use as the basis of his work on semiconductors?

For discussion

Which pieces of equipment in your home do you think have resistors controlling the flow of electricity? You may like to look back at page 203 to help you answer.

4 Some bedside clocks have a display that glows dimly when the room is dark yet shines brightly if the room is light. How could an LDR be responsible? How may the LDR make it easier for you to get to sleep?

Diodes

A current can only flow through a diode in one direction. They are used to control the direction of flow of a current through complicated circuits, such as those used in a radio, which have components in series and parallel.

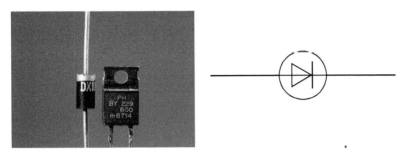

Figure B Two diodes and the symbol for a diode

5 Compare the actions of a resistor and a diode.

Diodes have a band marked at one end. When the diode is connected into a circuit the end with the band on must be connected to a wire coming from the negative terminal of the cell or battery for the current to flow. When a diode symbol is drawn in a circuit diagram the symbol should be drawn with the straight line facing the negative terminal of the current source.

A light-emitting diode is often referred to as an LED. In simple circuits we often use a lamp to show that a current is flowing. In electronic circuits an LED performs the same task more efficiently. An LED is a semiconductor diode, allowing a current to flow in only one direction through it, and it produces light. An LED can emit red, yellow or green light. The colour emitted depends on the semiconductor materials used to make the LED.

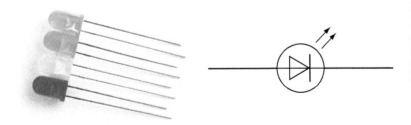

Figure C Four light-emitting diodes (LEDs) and their symbol

6 a) How is an LED similar to a lamp?
b) How is an LED different from a lamp?

◆ SUMMARY ◆

◆ A closed or completed circuit is needed for an electric current to flow (*see page 203*).

◆ When circuits are drawn, symbols are used for the parts or components (*see page 204*).

◆ The lining up of components in a row, end to end, is called arranging them in series (*see page 205*).

◆ In a parallel circuit, components are arranged side by side (*see page 211*).

◆ Materials that allow an electric current to pass through them are called conductors (*see page 205*).

◆ Materials that do not allow an electric current to pass through them are called insulators (*see page 205*).

◆ Resistors control the size of the electric current flowing in a circuit (*see page 206*).

◆ A buzzer vibrates to produce sound when an electric current passes through it (*see page 207*).

◆ Fuses and circuit breakers prevent circuits carrying too great an electric current (*see page 208*).

◆ The material through which a current flows offers some resistance to the moving electrons (*see page 208*).

◆ An ammeter measures the rate of flow of electrons, or current (*see page 209*).

◆ Lamps can be used to estimate the size of a current in a circuit (*see page 211*).

◆ The volt is the unit used to measure the potential difference between two points (*see page 212*).

◆ A voltmeter is used to measure the voltage, or potential difference, between two points in a circuit (*see page 213*).

End of chapter questions

1 Make a list of the components in each of the three circuits shown in Figure 17.20.

a) 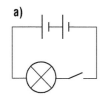   **b)** 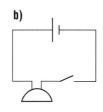   **c)**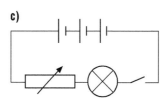

Figure 17.20

2 A circuit contains a cell, an LDR, an LED and a switch. When the switch is closed and the circuit is left in daylight the LED glows, but when the closed circuit is left in the dark the LED no longer glows. Explain what is happening in the circuit in both the light and the dark.

◆ Heat and internal energy
◆ Measuring the amount of heat energy
◆ Conduction
◆ Convection
◆ Radiation
◆ Evaporation

Heat and internal energy

The 'heat' in a substance is really a measure of the total kinetic energy of the atoms and molecules of the substance, due to its internal energy. The total amount of heat in a substance is related to its mass. A large mass of a substance holds a larger amount of heat – it has more internal energy – than a smaller mass. For example, if $100 \, cm^3$ of water is heated in a beaker with a Bunsen burner on a roaring flame it will take less time to reach $100\,°C$ than $200 \, cm^3$ of water would take, because it has a smaller mass (Figure 18.1).

Figure 18.1 When heating two masses of water, more heat energy needs to be supplied to the larger mass to reach the same temperature.

When a substance is heated, the (thermal) energy supplied increases the internal kinetic energy, which means the atoms and molecules in the substance move faster and further. If the temperature of the substance is taken with a thermometer, kinetic energy from the substance passes to the atoms or molecules from which the thermometer liquid is made and causes them to move faster too. This leads to an expansion of the liquid in the thermometer tube.

The thermometer measures the (average) kinetic energy of the particles hitting the bulb and not the total kinetic energy of all the particles in the substance.

1 Why does it take a full kettle longer to boil than a half-full kettle?

2 Which do you think contains more internal energy, a teaspoon of boiling water or a pan full of water at 50 °C?

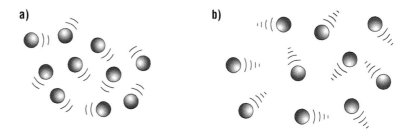

Figure 18.2 Particles in motion in **a)** a cool substance and **b)** a hot substance

Measuring the amount of heat energy

The amount of heat (thermal) energy given to a substance can be measured by heating the substance with an electric heater. The quantity of electrical energy used can be measured using a joulemeter and this equals the amount of heat (thermal) energy supplied.

The equipment in Figure 18.3 can be used to compare how the heat supplied to a liquid or solid affects its

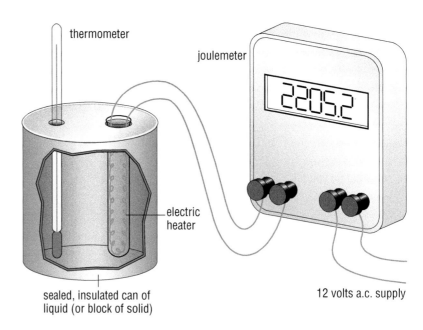

Figure 18.3 Measuring heat energy supplied with a joulemeter

temperature. It is found that some substances, such as water, take up large amounts of heat energy but their temperature only rises a few degrees, while the same mass of other substances needs only a small amount of heat energy to raise their temperature by the same amount.

How heat energy travels

There are three ways in which heat energy can travel. They are conduction, convection and radiation. Together they are known as thermal energy transfer.

Conduction

In **conduction**, the heat energy is passed from one particle of a material to the next particle. For example, when a metal pan of water is put on a hot plate of a cooker the atoms in the metal close to the hot plate receive heat energy and vibrate more vigorously. They knock against the atoms a little further into the bottom of the pan and make them vibrate more strongly too. These atoms knock against other atoms a little further up and the kinetic energy is passed on. Eventually the inner surface of the pan, which is next to the water, becomes hot too (Figure 18.4).

Conduction can occur easily in solids, less easily in liquids but hardly at all in gases because the gas atoms are too far apart to affect each other. It cannot occur in a vacuum, such as outer space, where there are no particles to pass on the heat energy. Conduction is fastest in metals because they have electrons that are free to move. When a metal is heated the electrons in that part move about faster and pass on heat energy to nearby electrons and atoms, so that the heat energy spreads quickly through to cooler parts of the metal.

Materials that allow heat to pass through them easily are called conductors of heat. Materials that do not allow heat to pass through them easily are called insulators.

Insulators are useful in reducing the loss of heat energy. For example, a thick woollen pullover is a good insulator because the wool fibres trap air, which is a poor conductor. This means that the pullover keeps in your body heat in cold conditions. Insulation materials are used in attics and lofts to reduce the escape of heat through the roofs of houses. These materials are not only poor conductors of heat themselves, but they also trap air, which is an excellent insulator, to reduce conduction further.

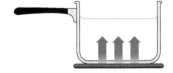

Figure 18.4 The conduction of heat through the bottom of a pan

3 A metal rod has drawing pins stuck to it with wax and is heated at one end as shown in Figure 18.5.

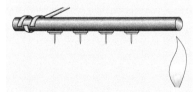

Figure 18.5

a) What do you think will happen in this experiment? Explain your answer.

b) How could this experiment be adapted to compare the conducting properties of different materials?

4 Imagine that a football represents heat energy and football players represent particles in a material. Which of the following events is like conduction, and which is like convection? Explain your answers.

a) The players pass the ball to each other to move it up the field.

b) A defender receives the ball and runs up field with it into an attacking position.

5 When coal burns, particles of soot rise up above the fire and make smoke. Why doesn't the smoke move along the ground?

6 a) The temperature of the land surface is higher than the temperature of the sea surface during the day. Use the ideas of convection currents to suggest what happens to air above the land and above the sea.

b) Which way do you think the wind will blow across the promenade in Figure 18.7, during the day? Explain your answer.

Figure 18.7

c) At night the land surface is cooler than the sea surface. Does this affect the wind direction? Explain your answer.

Convection

In **convection**, the heat energy is carried away by the particles of the material changing position. For example, the water next to the hot surface at the bottom of the pan receives heat from the metal. The molecules of water next to the metal move faster and further apart as their kinetic energy increases. This makes the water next to the pan bottom less dense than the water above it and the warm water rises. Cooler water from above moves in to take the place of the rising warmer water. The cool water is also warmed and rises. It is replaced by yet more cool water and convection currents are set up as shown in Figure 18.6.

Figure 18.6 The convection currents in a pan of water heated from below

Convection can only occur in liquids and gases. It cannot occur in solids where the particles are not free to move about, nor in a vacuum such as outer space.

Radiation

Energy can travel through air or through a vacuum as electromagnetic waves by **radiation**. For example, as the pan of water gets hotter you can put your hand near its side and feel the heat on your skin even though you are not touching the metal. The sides of the pan are radiating infrared waves. These carry the heat energy from the surface of the pan to the surface of your skin, which is warmed by them.

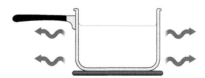

Figure 18.8 Heat radiation from a hot pan

All objects radiate infrared, but the hotter the object the more infrared energy it radiates and the shorter the wavelength of the waves.

Some infrared radiation can pass through certain solids such as glass. For example, the infrared radiation from the Sun can pass through glass in a greenhouse but the (longer wavelength) infrared radiation from the ground and the plants inside the greenhouse cannot pass back out through the glass. This infrared radiation is trapped and warms the contents of the greenhouse.

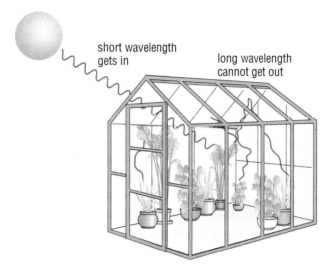

short wavelength gets in

long wavelength cannot get out

Figure 18.9 A greenhouse traps infrared energy

7 In question **4** the energy was transferred by particles (players). How is the transfer of heat energy by radiation different?

8 How does the type of surface of an object affect the way it radiates and absorbs heat energy?

The type of surface affects the amount of heat energy radiated from an object in a given time. Darker colours radiate energy more rapidly than lighter colours, and black surfaces radiate the most rapidly. The surfaces that radiate the energy least rapidly are light, shiny surfaces like polished metal.

The type of surface also affects the amount of radiated energy absorbed by the surface in a given time. For example, a light, shiny surface absorbs energy the least rapidly, while a dull, black surface absorbs energy the most rapidly.

The Thermos flask

Sir James Dewar (1842–1923) studied how gases could be turned into liquids. Liquid oxygen has a boiling point of −182.9 °C. When the liquid oxygen was made it needed to be stored in a container that would prevent heat from the surroundings entering and causing the liquid to boil. In 1892, Dewar invented a flask called a Dewar flask that allowed him to keep the liquid oxygen cool for his experiments. The Dewar flask is now more widely known as the Thermos flask (Figure A) and is used mainly to keep drinks hot.

The walls of the flask are made of glass, which is a poor conductor of heat, and are separated by a vacuum. The glass walls themselves have shiny surfaces. The surface of the inner wall radiates very little heat and the surface of the outer wall absorbs very little of the heat that is radiated from the inner wall. The cork supports are poor conductors of heat and the stopper prevents heat being lost by convection and evaporation in the air above the surface of the liquid.

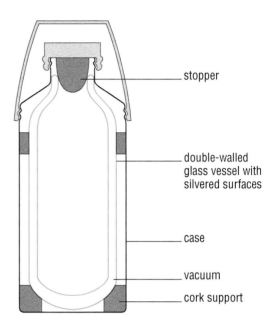

Figure A Structure of a Thermos flask

stopper

double-walled glass vessel with silvered surfaces

case

vacuum

cork support

1 What creative thought did Dewar have when thinking about studying liquid oxygen?

2 What evidence from the work of others did he use in designing his flask?

3 Which forms of energy transfer does the vacuum prevent?

4 How would the efficiency of the flask be affected if the walls were painted black? Explain your answer.

5 How could a warm liquid lose heat if the stopper was removed? Explain your answer.

Evaporation

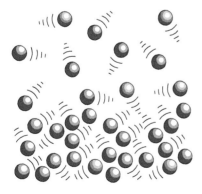

Figure 18.10 Evaporation

The particles in a liquid have different amounts of energy. The particles with the most energy move the fastest. High energy liquid particles near the surface move so fast that they can break through the surface and escape into the air and form a gas. This is called **evaporation**. When these particles leave, the amount of energy remaining in the liquid is reduced and the temperature of the liquid drops making it cooler.

The human body must be kept at 37 °C for good health. When we exercise, our muscles release some energy as heat and the body temperature rises. This causes the sweat glands in the skin to release sweat onto the skin surface. As the sweat evaporates its temperature falls and this draws heat from the body to the surface and cools the body down. You can feel the effect of evaporation cooling your skin by moistening a finger and blowing on it.

Evaporation is also used to keep refrigerators and freezers cool. A gaseous substance called a refrigerant is squashed in a compressor. The increased pressure moves the particles together so much that the refrigerant turns into a liquid. This is released into tubes in the cold compartment of a refrigerator or freezer. As it is released its pressure drops rapidly and this causes a huge number of particles to evaporate and form a gas. The temperature of the remaining liquid falls and draws heat out of the air in the compartment and reduces its temperature too. The refrigerant then leaves the compartment and returns to the compressor where it is squashed again before returning to the compartment and cooling it some more. This cycle of events continues until the required temperature of the refrigerator or freezer is reached.

◆ SUMMARY ◆

◆ When a substance is heated its internal energy increases and its particles move faster (*see page 217*).
◆ A thermometer measures the average kinetic energy of the particles hitting the thermometer bulb (*see page 218*).
◆ Heat energy supplied can be measured with a joulemeter (*see page 218*).
◆ Heat energy is passed from particle to particle by conduction (*see page 219*).
◆ Materials can be divided into conductors and insulators according to how easily heat passes through them (*see page 219*).
◆ Heat energy is carried by moving particles in a convection current (*see page 220*).
◆ Heat energy is carried by electromagnetic waves in radiation (*see page 220*).
◆ Evaporation reduces the amount of energy in the particles in a liquid and this brings about cooling (*see page 222*).

End of chapter questions

An investigation was carried out to see if useful amounts of heat energy from the Sun could be trapped in trays of water. Three metal trays were used. Each one was 25 cm long, 20 cm wide and 5 cm deep and was filled with 1500 cm³ of water.

- Tray 1 had a glass plate cover and the water in it was untreated.
- Tray 2 had some black ink added to the water before the glass plate cover was put over it.
- Tray 3 had some black ink added to the water before the glass plate cover was put over it, then the sides and base were packed with vermiculite – a sponge-like, rocky material.

The trays were exposed to sunlight during the day for 7 hours and the air temperature and the temperature of the water in each tray were taken every hour. Table 18.1 shows the data that were collected.

Table 18.1

Time	Air temperature/°C	Water temperature/°C		
		colourless (without insulation)	black (without insulation)	black (with insulation)
11.15a.m.	18.7	17.5	17.5	17.5
12.15p.m.	19.0	18.9	20.0	20.4
1.15p.m.	20.5	22.4	25.1	26.0
2.15p.m.	18.3	21.0	23.0	24.9
3.15p.m.	18.9	21.5	23.4	25.6
4.15p.m.	20.6	24.5	28.3	29.1
5.15p.m.	20.2	28.5	31.8	33.2
6.15p.m.	19.2	26.0	28.1	31.5

1 Did the water fill the trays to the top? Explain your answer.

2 Plot lines of the data for the air temperature and the temperature of each tray, on the same graph.

3 Compare the lines of the data on the graph you have drawn.

4 What was the purpose of the ink and the vermiculite? Explain your answers.

5 What is the maximum temperature rise? When was it achieved and in which tray?

6 What can you conclude from this investigation?

19 World energy needs

For discussion

What do people think will happen when the energy crisis comes? How do they think they could try and avoid the crisis happening?

We are moving towards a time when there will be an energy crisis. A crisis is a time of danger or of great difficulty. We are in danger of not having enough energy to meet our present needs. Try the *For discussion* activity before you read on.

Energy pathways on Earth

The Sun sends out a huge amount of energy into space, in all directions, every second. Only a tiny fraction of this energy reaches the Earth. Even so, this amount of energy from the Sun is large in our terms on Earth. It would take 173 million large power stations to produce the same amount of energy that we receive from the Sun each second. However, not all of the Sun's energy that reaches the Earth can be useful to us.

Figure 19.1 shows a Sankey diagram about energy pathways on Earth. It features energy from the Sun, geothermal energy and energy from the tides. It does not include nuclear energy, which is described on pages 232–234.

For discussion

What are the sources of the energy that reaches the Earth's surface? If each one in turn was no longer available, could we survive?

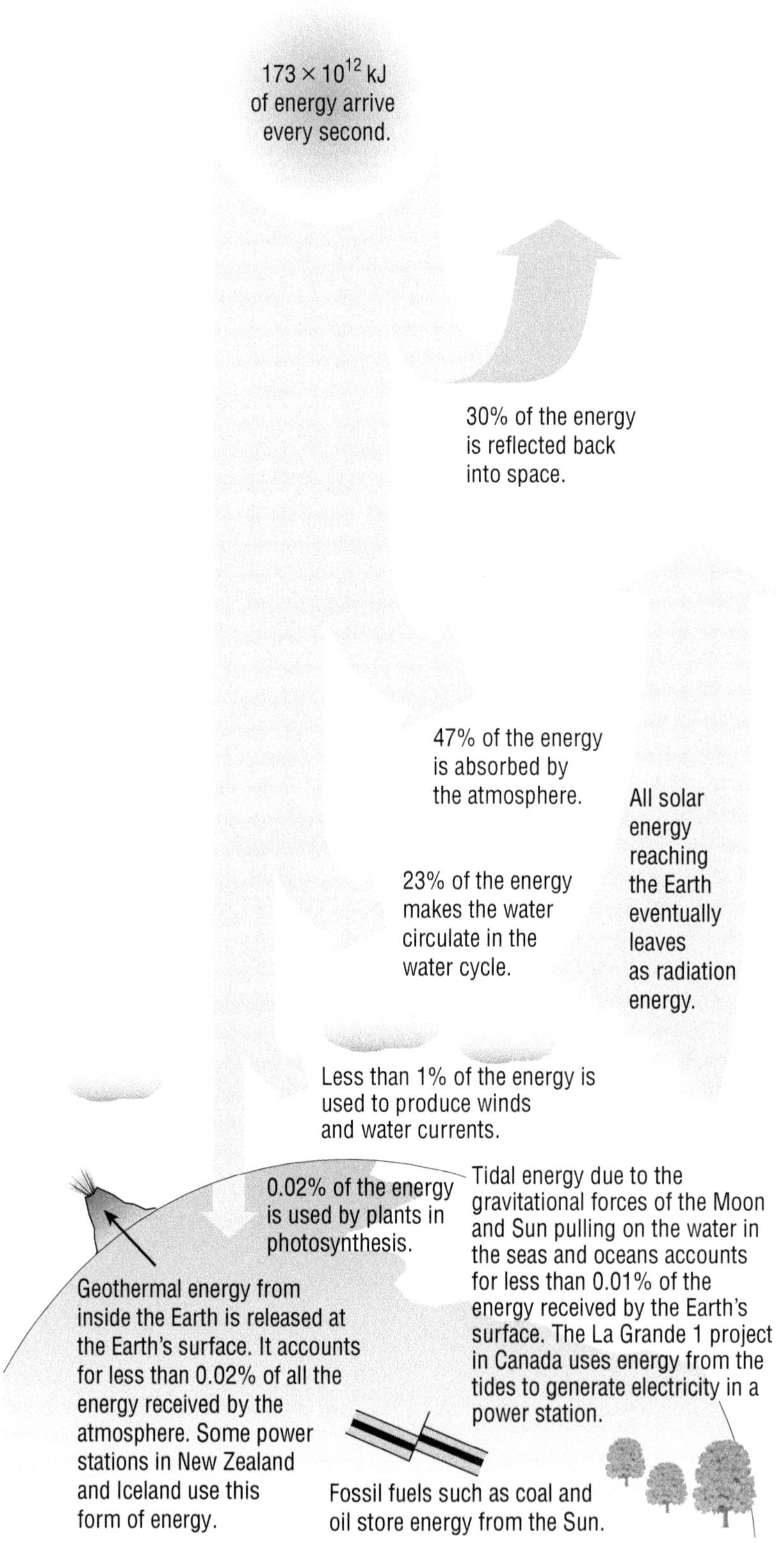

173 × 10¹² kJ of energy arrive every second.

30% of the energy is reflected back into space.

47% of the energy is absorbed by the atmosphere.

All solar energy reaching the Earth eventually leaves as radiation energy.

23% of the energy makes the water circulate in the water cycle.

Less than 1% of the energy is used to produce winds and water currents.

0.02% of the energy is used by plants in photosynthesis.

Tidal energy due to the gravitational forces of the Moon and Sun pulling on the water in the seas and oceans accounts for less than 0.01% of the energy received by the Earth's surface. The La Grande 1 project in Canada uses energy from the tides to generate electricity in a power station.

Geothermal energy from inside the Earth is released at the Earth's surface. It accounts for less than 0.02% of all the energy received by the atmosphere. Some power stations in New Zealand and Iceland use this form of energy.

Fossil fuels such as coal and oil store energy from the Sun.

Figure 19.1 Sankey diagram showing the flow of energy to and from the Earth's surface

The water cycle

Nearly a quarter of the Sun's energy drives the water cycle on Earth. It heats the surface of the water in the oceans and seas causing large amounts of water to evaporate. The water vapour passes into the air and eventually condenses on dust. The tiny droplets of condensed water form clouds. Winds push the clouds along and when they reach cool regions such as high mountains, larger droplets form and rain is released. The rainwater forms rivers, which lead back to the seas.

Winds and waves

When the Sun's rays heat the ground some of the heat is passed into the air. The warm air rises and is replaced by cooler air. This movement of the air generates winds. Some winds blow on the surface of seas and cause waves to form. The movement of the air and the up-and-down movement of the waves can be used to generate electrical energy (Figure 19.10, page 236).

Tidal and geothermal energy

There are two energy paths which do not begin with the Sun. One energy path begins with the movement energy in the tides as they rise and fall due to the pull of gravity of the Moon and Sun on the waters of the seas and oceans. The second energy path begins with the heat energy released by radioactive materials (page 232) inside the Earth.

Fossil fuels

Only a very small amount of the Sun's energy is trapped by plants and used to make food. The food is used by plants and animals. When plants and animals die, their bodies decompose and the energy that they possessed passes into decomposing organisms, such as bacteria, and eventually passes out into the air and space as heat. In the past there were regions in the world where this did not happen. The energy in the bodies of ancient trees and marine organisms was trapped underground in their bodies. These organisms did not decay and fossils fuels (coal, oil and gas) were formed.

1 How much of the Sun's energy is reflected back into space or absorbed by the atmosphere?
2 Make a drawing of the water cycle from the information in the paragraph on the right.
3 The water at the head of a river is much higher than the water at sea level. What kind of energy does it possess?
4 How may the energy possessed by moving water be used to provide us with energy we can use?

5 What is the name of the process by which plants make food from the energy in sunlight?

Figure 19.2 These bottles contain samples of oil from different parts of the world.

From one crisis to another

The first fuel that humans used was wood. There was plenty of it around and the trees grew back at the rate at which people used them. In time, in some countries, the number of people increased and the demand for fuel increased as well. The trees could not provide enough energy to meet the needs of the people and an energy crisis loomed. Also the destruction of trees led to the washing away of the soil in floods leaving ground on which few plants would grow.

The possible energy crisis due to tree destruction was averted by the discovery of fossil fuels – coal, oil and gas. They are the fuels on which much of the world depends today. The problem with a fossil fuel is that it takes millions of years to develop. It cannot form at a rate matching that at which we use it up, so eventually it will all be gone.

The problems with fossil fuels

It is quite easy for people to collect wood for fuel. They can just pick up dead branches or chop down a tree. The fuel has then only to be transported, possibly a short distance, to the place where it will be used. When people use fossil fuels there are more problems to overcome.

Extracting coal

Coal forms in layers below the ground. The layers are called seams. If the coal seam is near the surface of the ground, the soil and rocks above it are removed and an open-cast mine is formed (Figure 19.3). The coal seam is removed and the rocks and soil are replaced.

Figure 19.3 Open-cast mining

If the coal seams are deep, shafts are sunk to meet them and then tunnels are dug into them (Figure 19.4). Coal is dug out of the seam at a wall called the coal face. It is transported back down the tunnel on trucks or conveyer belts and brought to the surface in lifts. Miners travel in and out of the mines in lifts too.

6 Look at Figure 19.4 and imagine that you are about to set up a mine to extract coal from deep underground. What will you have to spend money on before you can begin extracting the coal?

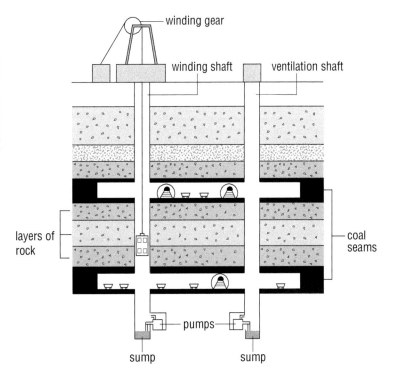

Figure 19.4 Mining deep coal seams

Two dangers in the mine are firedamp and flooding. Methane gas escapes from the coal as it is being mined and mixes with the air to form firedamp. If this mixture is lit it explodes. The build up of firedamp in a mine is avoided by sinking ventilation shafts and driving a current of air through the shafts and tunnels. Flooding is prevented by building sumps (pits into which water drains) at the bottom of shafts and pumping the water that collects there to the surface.

Extracting oil and gas

Oil and gas collect underground in some places where a porous rock is sandwiched between two layers of non-porous rock. A narrow shaft is drilled through the upper layer of non-porous rock to reach the layer of porous rock. The gas and oil pass up the shaft and are carried away along pipes to be stored and transported.

Some gas and oil is present under the land in certain countries while in other places it is present under the seabed. Extracting oil from under the sea requires the building of a platform on which a drill can be set up. The platform also provides a home for the people working on drilling and extracting the fuels.

7 Imagine that you have been given the task of extracting and selling oil from an oil well. You have to set up the oil well and transport the oil to the buyers. There are two possible oil reserves you could exploit – one is under the land, and the other lies offshore, under the seabed. Which well do you think will be cheaper to set up and run? Explain your answer.

Figure 19.5 *A typical offshore platform, above an oil well under the sea*

The price of fuel

The price of fuel is related to the cost of extracting it. It is cheaper to extract a fuel like coal when it is near the surface as the shafts are short and less energy is used on lifting the fuel to the surface. When the coal seams near the surface are used up, deeper seams have to be mined. This means making deeper shafts and using more energy lifting the fuel to the surface. This makes the fuel more expensive. In time it can become too expensive compared to other fuels and the mine is closed.

Mines and wells are set up close to places where the fuels are used or where they can be transported cheaply to the people who need them. As these mines and wells are used up, other places where the fuels can be found may be used. These places may be more difficult to reach and increase the costs of setting up mines and wells and of transporting the fuels.

Fossil fuels and the environment

When fossil fuels burn they produce carbon dioxide. As we use large amounts of fossil fuels, large amounts of carbon dioxide are produced. This gas mixes with the gases in the air. The presence of carbon dioxide in the air makes the atmosphere behave like the glass in a greenhouse (page 221). As extra carbon dioxide is produced this effect is increased. This causes the temperature of the atmosphere to rise and results in an increase in global warming.

8 Your oil well, which is on land, is running dry and you get the chance to set up a new well 100 km further away over some rugged hills where there are no roads. What extra expenses would you have to pay to extract this oil? How would the price of oil affect your decision to extract the oil?

Figure 19.6 A rise in temperature, possibly due to global warming, has killed this tree and others like it in the area.

Figure 19.7 Acid rain has damaged this stone statue

Fossil fuels contain an element called sulfur. When the fuels are burnt the sulfur combines with oxygen to form sulfur dioxide. This reacts with oxygen and water vapour in the air and forms sulfuric acid. It may fall to the ground as acid rain. When acid rain passes into the soil it removes some of the minerals from the soil that plants need for healthy growth. The plants' growth slows and they become stunted or even die. When the acid rain reaches rivers and lakes it causes the acidity of the water to increase. Many aquatic organisms cannot adapt to the increase in acidity and die. In towns, the ornamental stonework on buildings may be destroyed by acid rain as shown in Figure 19.7.

For discussion

There have been periods in the Earth's history when conditions were warmer than today. At those times fossil fuels were not being burnt. Some scientists believe that global warming now is not due to fossil fuels. Others believe that fossil fuels are the major cause. Research books, newspapers and the internet for different views. Look for evidence being used to support a view. Consider the evidence of both views. Do they balance or does one set of evidence appear stronger than the other? What is the view that is most common today? Should people cut down on the use of fossil fuels?

Non-renewable fuels

Fossil fuels are sources of **non-renewable** energy. They cannot be replaced because the rate at which they form is much lower than the rate at which they are being used up. Radioactive materials are also a non-renewable energy source.

Radioactive materials

Almost all the elements forming the matter from which the Earth is made have been made by nuclear fusion in stars in the distant past. Some elements are unstable and 'decay' in nuclear reactions to form more stable elements. These unstable elements are called **radioactive elements**. When they undergo radioactive decay they release particles and energy.

Using nuclear fission

Albert Einstein (1879–1955), a German-born self-taught physicist, began by studying the work of other scientists. Among other enlightened theories, he produced an equation that linked mass and energy. The equation is:

$E = mc^2$

where E is energy, m is mass and c is the velocity of light.

This equation showed that mass is a store of energy, which can be released when matter is destroyed. This idea successfully explained the phenomenon of radioactive materials by showing that they produced energy in the form of radiation because part of their mass was destroyed. Otto Hahn (1879–1968), a German chemist, and Lise Meitner (1878–1968), an Austrian physicist, studied how radioactive materials decay and in 1939 Meitner described how uranium atoms break in half.

Leo Szilard (1898–1964), a Hungarian-born physicist who studied atomic nuclei, had an idea of how the break-up of one atom may cause the break-up of surrounding atoms and lead to a chain reaction. When Szilard heard of Hahn's and Meitner's work he realised that uranium could be investigated to see if a chain reaction would be set up (Figure D). If such a reaction could take place then a large amount of energy could be released quickly.

Szilard was drawing up his plans and predictions in the United States during the Second World War. He realised that the chain reaction could be used to release energy for a bomb. He and other scientists, including Einstein, informed the United States President of the explosive power of uranium. This led to the development of nuclear bombs. Szilard and many scientists believed the bombs were too devastating to be used like other bombs and should only be demonstrated in an uninhabited part of the world, to show the enemy the power that was now set against them.

However, politicians in charge of the armed forces and some scientists believed it right to use the nuclear bombs like other bombs, and two were dropped on Japan, with catastrophic results.

After the Second World War scientists began investigating ways of using nuclear energy for peaceful purposes. This meant that the energy released in the chain reaction had to be released more slowly than in a bomb.

Figure A Albert Einstein put forward many enlightened theories.

Figure B Lise Meitner studied radioactive decay.

Figure C Devastation caused by a nuclear explosion in Hiroshima, Japan, 1945

1 Why do radioactive materials release energy?
2 Why do you think the scientists first considered making nuclear bombs instead of reactors for power stations?
3 What could be the consequences of an explosion at a nuclear reactor?
4 What are the advantages and disadvantages of using nuclear fuel for generating electricity?

5 What was Szilard's creative thought?
6 What evidence about nuclear fission did he use?

The nuclear reactor was developed to release energy in sufficient amounts to heat water and produce steam. This could then be used to spin a turbine in a generator to produce electricity. Today there are about 400 nuclear reactors operating in more than 30 countries.

Nuclear reactors are built and operated to very strict safety rules to prevent them overheating and exploding. They do not produce sulfur dioxide and carbon dioxide like the power stations using fossil fuels, but the nuclear wastes that they produce have to be stored for thousands of years while they decay to harmless materials.

> For discussion
>
> **'The study of radioactive materials has been a great benefit to humankind.'**
>
> **Discuss this statement.**

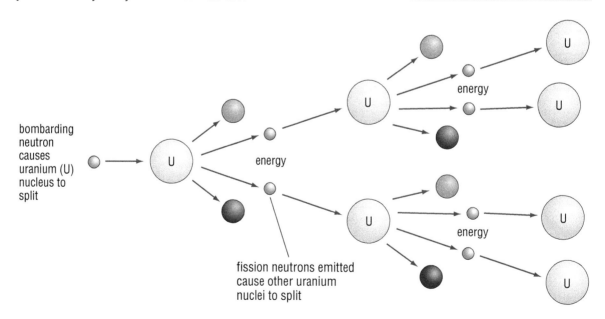

bombarding neutron causes uranium (U) nucleus to split

energy

fission neutrons emitted cause other uranium nuclei to split

energy

Figure D As one uranium atom splits, it fires out neutrons, which trigger other uranium atoms to split, so causing a chain reaction.

Renewable energy sources

Renewable energy resources can be replaced in a short time and so are not used up.

Solar energy

We can use two forms of solar energy as sources of renewable energy. Heat energy can be trapped by solar panels (Figure 19.8) and used to heat buildings. Light energy can also be converted into electrical energy by solar cells (Figure 19.9).

For discussion

Can a house be designed that would get all its energy from solar panels and solar cells? In the design there can be batteries to store electrical energy.

Figure 19.8 These roofs in Sweden are fitted with solar panels.

For discussion

Can a small town be designed to run on energy from solar panels and solar cells?

Figure 19.9 Arrays of solar cells project from the International Space Station to collect some of the Sun's energy.

Waves as energy sources

Waves move up and down as they pass by. Machines are being developed to convert the up-and-down motion of the wave into a circular motion, which can turn the shaft of a turbine and generate electricity. The machines need to be moored in places where the waves are frequent but not too strong to destroy them (see Figure 19.10).

9 Draw an energy transfer diagram for a generator using the energy in a wave.

10 What expenses would have to be met to set up machines to generate electricity from waves and transport the electricity to the shore?

Figure 19.10 This machine converts the energy in the up-and-down motion of passing waves into a turning motion required to make an electric generator work.

Rivers as energy sources

The kinetic energy in moving water in a river is a source of renewable energy. However, the flow of water for providing useful energy for people needs to be controlled. If a turbine was simply lowered into a river and left there, it would spin when the river flowed normally, it would hardly turn at all when the river had little water in it and it might be washed away when the river was in flood. This was known by the early engineers who set up water mills.

They also set up a means of controlling the flow over the water wheel. The water was released steadily from a dam with a lodge or lake behind it.

Today dams are built to store huge amounts of water to turn the turbines in hydroelectric power stations (Figure 19.11).

For discussion

Large dams are built to provide the energy for large power stations. When a large dam is built a large reservoir forms behind it, as Figure 19.11 shows. Often there are villages on the land that is to be flooded. Imagine you have to explain to the leader of a village that the people in the village will have to move. You have to make the movement of villagers out of the area as peaceful as possible. You can offer them incentives to move such as money and land but you must try and move them as cheaply as possible. You may like to develop this activity into someone taking the role of the village leader and others taking the role of villagers while you and a few others take the roles of people representing the company building the dam and power station.

Figure 19.11 Water released from this dam is used to turn turbines connected to generators in a hydroelectric power station.

Wind as a source of energy

Just as there has to be a steady flow of water for harnessing energy from the current, so there has to be a steady flow of wind for a wind turbine to work efficiently.

Obstacles such as buildings can cause turbulence in the air when the wind passes over them and if the turbine is sited near a building it may not get a steady flow of the wind to keep it turning smoothly.

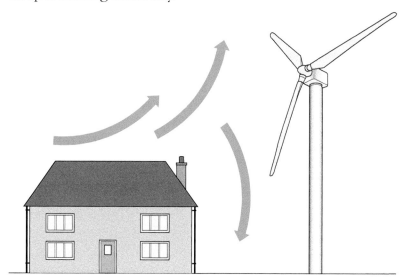

Figure 19.12 House near turbine

When wind blows against a cliff face, turbulence develops at the top. If a turbine is sited at a clifftop it will not turn smoothly.

Figure 19.13 Turbine on clifftop

A hill that rises steadily to a rounded hilltop, then sinks slowly on the other side, is a good site for a turbine. It should be set up at the top of the hill. The wind here will move smoothly over the hilltop like the air on an aircraft wing and keep the turbine turning steadily.

However, wind is not the only consideration in siting a wind farm. Turbines also make a noise. As the strength of the sound dies away with the increasing distance from the source of the sound, turbines need to be sited away from habitations.

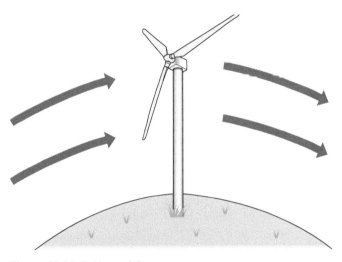

Figure 19.14 Turbine on hilltop

For discussion

Imagine you live in a town on the coast. Behind the town are two hills. One hill has no mountains behind it and no one lives there. It is covered in small plants and low bushes and is the home of many wild birds. There are a few paths across it for hikers to use. The second hill has mountains behind it and there is a town at its foot. The council (the elected people who run the town) decide that they could cut down on the use of coal and oil to generate electricity by setting up a wind farm. The problem is where to build it. It could be on the hill without the mountains behind where it could be easily seen from the town, or on the hill with the mountains behind it where the turbines would be less easily seen but their noise might disturb the people in the town. Or it could be built out at sea where it would not be seen but there would be more expense in bringing the electricity to the town.

When issues such as this – selecting a site for a wind farm – arise, the media (newspapers, radio and television) run reports and articles about them. They often take one view in order to try and influence the decisions of the people affected by the issue. Imagine that you are a science reporter who is going to write an article giving a balanced view, setting out the advantages and disadvantages of different sites for the wind farm. Research your article by reading more about renewable and non-renewable energy. Write your article and see if your friends think it is balanced or if you seem to favour one site more than the other.

◆ SUMMARY ◆

- ◆ Energy pathways on Earth can be shown in a Sankey diagram (*see page 226*).
- ◆ Some of the Sun's heat energy drives the Earth's water cycle (*see page 227*).
- ◆ Some of the Sun's heat energy generates the winds and waves (*see page 227*).
- ◆ Fossil fuels store energy from the Sun (*see page 227*).
- ◆ The world today depends on fossil fuels for energy (*see page 228*).
- ◆ There are problems with the extraction of coal, oil and gas from the ground (*see page 228*).
- ◆ The price of fuel is related to the cost of extracting it (*see page 231*).
- ◆ Burning fossil fuels can cause damage to the environment (*see page 231*).
- ◆ Radioactive elements release energy as they decay (*see page 232*).
- ◆ Energy resources which can be replaced are called renewable energy sources (*see page 235*).

End of chapter questions

1 What energy sources are used in your region?

2 What impact do they make on the environment? For example, have valleys been flooded? Is there air pollution?

3 **a)** If you were to introduce alternative energy sources to your area what would you choose?

b) How do you think people in your area would react to your ideas?

Theories of everything

I hope that after three years of studying biology, chemistry and physics as part of the *Cambridge Checkpoint Science* course you have a better understanding of our world. At times in this book there have been references to how two or all three of the subjects are linked together and you may like to take this a stage further and try and explain your surroundings in scientific terms.

For example, in this picture you could begin with the logs on the fire and explain that photosynthesis many years ago stored the Sun's energy in a plant, which in turn stored some of this energy in wood. The burning of the wood is releasing this energy as heat, which is passing out as radiation to warm the faces of the people round the fire, and passing by conduction through the cooking pot and by convection in the liquid in the food.

For discussion

What other relationships can you describe scientifically in the picture? For example, what is the relationship between respiration in the people around the fire and the plants surrounding them?

Figure A A camp fire

If you continued describing the world in this way, you may feel that you are developing a 'theory of everything'. You would not be the first to do this. We have seen that the Ancient Greeks tried to explain their world from their observations and came up with a theory of everything based on what they called the elements – earth, fire, air and water. They thought that everything was made from these elements. This theory was popular for many centuries. Democritus's ideas about atoms as a theory of everything did not become popular until the work of John Dalton and others.

Theories of everything try to link up everything in a simple way. When Newton, from his studies on gravity, set out his law of universal gravitation it brought together the work of Galileo on how things fall, the work of Kepler on how planets move in their orbits, and observations of how tides are related to the phases of the Moon.

The work of scientists in the 19th century on electricity and magnetism led to the discovery that they were linked as electromagnetic waves such as light and heat. Later scientists focused on the structure of the atom again as we have seen with Thomson and Rutherford. The evidence of their studies stimulated scientists to delve deeper into atomic structure using apparatus called bubble chambers. These allowed the tracks of subatomic particles to be seen and photographed.

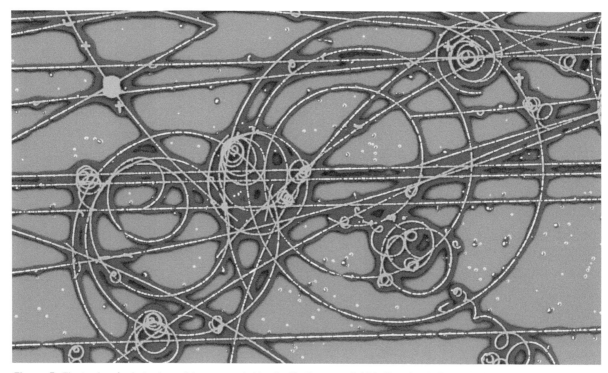

Figure B The tracks of subatomic particles as revealed by the Big European Bubble Chamber in Switzerland

The data produced by bubble chambers stimulated further interest and a huge piece of apparatus called the Large Hadron Collider, a high-energy particle accelerator, was built in a vast circular tunnel 175 m beneath the ground near Geneva in Switzerland. Work using this apparatus has shown that subatomic particles are held together by forces in the nucleus. In the future some scientists believe that they will be able to construct a theory of everything that links these forces with gravity and electromagnetism.

For the present, research goes on in many areas of science to make discoveries about our world and, where possible, to apply the discoveries to ways of helping us survive on the planet. You may feel that after your three years studying the *Cambridge Checkpoint Science* course that you might like to be part of this action, take further steps along the way to becoming a scientist and, as I hope you have done here, enjoy discovering more about our world and its place in the universe.

For discussion

How would you like to celebrate the completion of your *Cambridge Checkpoint Science* course?

How about a party? Should it be indoors or outside? Should it have a theme – everyone coming dressed as atoms and molecules, or as scientists from the past perhaps? Or you could make the theme wide open to any science topic you have studied and be surprised at the range of ideas and costumes. Could you think of a way of having science-themed food or drink? What about choosing music and inventing a dance? Scientists use their creative thinking in other ways besides science, as we have seen in the studies of scientists past and present.

Index